AF791609

CLIMATE CHANGE AND ITS CAUSES, EFFECTS AND PREDICTION

CLIMATE CHANGE ADAPTATION PLAN OF THE U.S. ENVIRONMENTAL PROTECTION AGENCY

Climate Change and its Causes, Effects and Prediction

Additional books in this series can be found on Nova's website under the Series tab.

Additional e-books in this series can be found on Nova's website under the e-book tab.

CLIMATE CHANGE AND ITS CAUSES, EFFECTS AND PREDICTION

CLIMATE CHANGE ADAPTATION PLAN OF THE U.S. ENVIRONMENTAL PROTECTION AGENCY

CORY ABRAHAMS
EDITOR

nova publishers
New York

Copyright © 2015 by Nova Science Publishers, Inc.

All rights reserved. No part of this book may be reproduced, stored in a retrieval system or transmitted in any form or by any means: electronic, electrostatic, magnetic, tape, mechanical photocopying, recording or otherwise without the written permission of the Publisher.

For permission to use material from this book please contact us:
nova.main@novapublishers.com

NOTICE TO THE READER

The Publisher has taken reasonable care in the preparation of this book, but makes no expressed or implied warranty of any kind and assumes no responsibility for any errors or omissions. No liability is assumed for incidental or consequential damages in connection with or arising out of information contained in this book. The Publisher shall not be liable for any special, consequential, or exemplary damages resulting, in whole or in part, from the readers' use of, or reliance upon, this material. Any parts of this book based on government reports are so indicated and copyright is claimed for those parts to the extent applicable to compilations of such works.

Independent verification should be sought for any data, advice or recommendations contained in this book. In addition, no responsibility is assumed by the publisher for any injury and/or damage to persons or property arising from any methods, products, instructions, ideas or otherwise contained in this publication.

This publication is designed to provide accurate and authoritative information with regard to the subject matter covered herein. It is sold with the clear understanding that the Publisher is not engaged in rendering legal or any other professional services. If legal or any other expert assistance is required, the services of a competent person should be sought. FROM A DECLARATION OF PARTICIPANTS JOINTLY ADOPTED BY A COMMITTEE OF THE AMERICAN BAR ASSOCIATION AND A COMMITTEE OF PUBLISHERS.

Additional color graphics may be available in the e-book version of this book.

Library of Congress Cataloging-in-Publication Data

ISBN: 978-1-63463-839-5

Published by Nova Science Publishers, Inc. † New York

CONTENTS

PREFACE

This book discusses the EPA's climate change adaptation plan, and discusses the impacts of climate change on the United States.

Chapter 1 - The U.S. Environmental Protection Agency (EPA) is committed to identifying and responding to the challenges that a changing climate poses to human health and the environment.

Scientific evidence demonstrates that the climate is changing at an increasingly rapid rate, outside the range to which society has adapted in the past. These changes can pose significant challenges to the EPA's ability to fulfill its mission. The EPA must adapt to climate change if it is to continue fulfilling its statutory, regulatory and programmatic requirements. The Agency is therefore anticipating and planning for future changes in climate to ensure it continues to fulfill its mission of protecting human health and the environment even as the climate changes.

In February 2013, the EPA released its draft *Climate Change Adaptation Plan* to the public for review and comment. The plan relies on peer-reviewed scientific information and expert judgment to identify vulnerabilities to EPA's mission and goals from climate change. The plan also presents 10 priority actions that EPA will take to ensure that its programs, policies, rules, and operations will remain effective under future climatic conditions. The priority placed on mainstreaming climate adaptation within EPA complements efforts to encourage and mainstream adaptation planning across the entire federal government.

Following completion of the draft *Climate Change Adaptation Plan*, each EPA National Environmental Program Office, all 10 Regional Offices, and several National Support Offices developed a *Climate Adaptation Implementation Plan* to provide more detail on how it will carry out the work called for in the agency-wide plan. Each *Implementation Plan* articulates how the office will integrate climate adaptation into its planning and work in a manner consistent and compatible with its goals and objectives.

Taken together, the *Implementation Plans* demonstrate how the EPA will attain the 10 agency-wide priorities presented in the *Climate Change Adaptation Plan.* A central element of all of EPA's plans is to build and strengthen its adaptive capacity and work with its partners to build capacity in states, tribes, and local communities. EPA will empower its staff and partners by increasing their awareness of ways that climate change may affect their ability to implement effective programs, and by providing them with the necessary data, information, and tools to integrate climate adaptation into their work.

Each Program and Regional Office's *Implementation Plan* contains an initial assessment of the implications of climate change for the organization's goals and objectives. These "program vulnerability assessments" are living documents that will be updated as needed to account for new knowledge, data, and scientific evidence about the impacts of climate change on EPA's mission. The plan then identifies specific priority actions that the office will take to begin addressing its vulnerabilities and mainstreaming climate change adaptation into its activities. Criteria for the selection of priorities are discussed. An emphasis is placed on protecting the most vulnerable people and places, on supporting the development of adaptive capacity in the tribes, and on identifying clear steps for ongoing collaboration with tribal governments.

Because EPA's Programs and Regions and partners will be learning by experience as they mainstream climate adaptation planning into their activities, it will be essential to evaluate their efforts in order to understand how well different approaches work and how they can be improved. Each *Implementation Plan* therefore includes a discussion of how the organization will regularly evaluate the effectiveness of its adaptation efforts and make adjustments where necessary.

The set of *Implementation Plans* are a sign of EPA's leadership and commitment to help build the nation's adaptive capacity that is so vital to the goal of protecting human health and the environment. Working with its partners, the Agency will help promote a healthy and prosperous nation that is resilient to a changing climate.

Chapter 2 - Report of the President Barack Obama on Executive Order 13653, Preparing the United States for the Impacts of Climate Change.

Chapter 3 - Report of an Executive Office of the President on The President's Climate Action Plan.

In: Climate Change Adaptation Plan …
Editor: Cory Abrahams
ISBN: 978-1-63463-839-5
© 2015 Nova Science Publishers, Inc.

Chapter 1

CLIMATE CHANGE ADAPTATION PLAN*

U.S. Environmental Protection Agency

PREFACE

The U.S. Environmental Protection Agency (EPA) is committed to identifying and responding to the challenges that a changing climate poses to human health and the environment.

Scientific evidence demonstrates that the climate is changing at an increasingly rapid rate, outside the range to which society has adapted in the past. These changes can pose significant challenges to the EPA's ability to fulfill its mission. The EPA must adapt to climate change if it is to continue fulfilling its statutory, regulatory and programmatic requirements. The Agency is therefore anticipating and planning for future changes in climate to ensure it continues to fulfill its mission of protecting human health and the environment even as the climate changes.

In February 2013, the EPA released its draft *Climate Change Adaptation Plan* to the public for review and comment. The plan relies on peer-reviewed scientific information and expert judgment to identify vulnerabilities to EPA's mission and goals from climate change. The plan also presents 10 priority actions that EPA will take to ensure that its programs, policies, rules, and

---

* This is an edited, reformatted and augmented version of a document produced by the Cross-EPA Work Group on Climate Change Adaptation Planning. The document, Publication Number: EPA 100-K-14-001, dated June 2014, was issued by the EPA October 31, 2014.

operations will remain effective under future climatic conditions. The priority placed on mainstreaming climate adaptation within EPA complements efforts to encourage and mainstream adaptation planning across the entire federal government.

Following completion of the draft *Climate Change Adaptation Plan*, each EPA National Environmental Program Office, all 10 Regional Offices, and several National Support Offices developed a *Climate Adaptation Implementation Plan* to provide more detail on how it will carry out the work called for in the agency-wide plan. Each *Implementation Plan* articulates how the office will integrate climate adaptation into its planning and work in a manner consistent and compatible with its goals and objectives.

Taken together, the *Implementation Plans* demonstrate how the EPA will attain the 10 agency-wide priorities presented in the *Climate Change Adaptation Plan.* A central element of all of EPA's plans is to build and strengthen its adaptive capacity and work with its partners to build capacity in states, tribes, and local communities. EPA will empower its staff and partners by increasing their awareness of ways that climate change may affect their ability to implement effective programs, and by providing them with the necessary data, information, and tools to integrate climate adaptation into their work.

Each Program and Regional Office's *Implementation Plan* contains an initial assessment of the implications of climate change for the organization's goals and objectives. These "program vulnerability assessments" are living documents that will be updated as needed to account for new knowledge, data, and scientific evidence about the impacts of climate change on EPA's mission. The plan then identifies specific priority actions that the office will take to begin addressing its vulnerabilities and mainstreaming climate change adaptation into its activities. Criteria for the selection of priorities are discussed. An emphasis is placed on protecting the most vulnerable people and places, on supporting the development of adaptive capacity in the tribes, and on identifying clear steps for ongoing collaboration with tribal governments.

Because EPA's Programs and Regions and partners will be learning by experience as they mainstream climate adaptation planning into their activities, it will be essential to evaluate their efforts in order to understand how well different approaches work and how they can be improved. Each *Implementation Plan* therefore includes a discussion of how the organization will regularly evaluate the effectiveness of its adaptation efforts and make adjustments where necessary.

The set of *Implementation Plans* are a sign of EPA's leadership and commitment to help build the nation's adaptive capacity that is so vital to the goal of protecting human health and the environment. Working with its partners, the Agency will help promote a healthy and prosperous nation that is resilient to a changing climate.

Bob Perciasepe
Deputy Administrator
September 2013

CLIMATE CHANGE ADAPTATION PLAN: EXECUTIVE SUMMARY

We live in a world in which the climate is changing at a rate faster than that which society has experienced in modern history. Because many of the environmental outcomes that EPA is working to attain (e.g., clean air, safe drinking water) are sensitive to changes in weather and climate, these changes are posing new challenges to EPA's ability to fulfill its mission of protecting human health and the environment.

To address these challenges, EPA has developed a *Climate Change Adaptation Plan.* The *Adaptation Plan* relies on peer-reviewed scientific information and expert judgment to begin to identify potential vulnerabilities to EPA's mission from climate change. The *Adaptation Plan* also presents priority actions the Agency will take to integrate climate adaptation planning into its programs, policies, rules, and operations to ensure they are effective in a changing climate. EPA's focus on climate adaptation is part of a larger federal effort to promote a healthy and prosperous nation that is resilient to a changing climate. EPA's *Climate Change Adaptation Plan* provides a road map and commitment from the Agency to addressing the direction provided in the President's *Climate Action Plan*, and Executive Order 13653, *Preparing the United States for the Impacts of Climate Change*.

Vision of the Future EPA

EPA's *Policy Statement on Climate-Change Adaptation*, issued in 2011, called for EPA to plan for future changes in climate and to mainstream considerations of climate change into its activities. As part of that effort, the *Policy Statement* called for the Agency to develop and implement a *Climate Change Adaptation Plan*. It also called for each EPA National Environmental Program Office and all of the Regional Offices to develop Implementation Plans to explain how they will carry out the work called for in the Agency-wide Plan. EPA made these Implementation Plans available as drafts for public comment late in 2013 and they have been revised with this Plan.

EPA's Climate Change Adaptation Plan

- EPA's Adaptation Plan identifies ways in which climate change could affect EPA's ability to fulfill its mission.
- EPA's Adaptation Plan describes priority actions that EPA will take to ensure that its programs, policies, rules, and operations will remain effective under future climatic conditions.

EPA's vision is for the Agency to continue to fulfill its mission of protecting human health and the environment even as the climate changes. EPA will build and strengthen its ***adaptive capacity*** and work with its partners to build capacity in states, tribes, and local communities. EPA will empower its staff and partners by increasing their awareness of ways that climate change may affect their ability to implement effective programs, and by providing them with data, information, and tools to integrate climate adaptation into their work.

Adaptive capacity is the ability of a human or natural system to adjust to climate change (including climate variability and extremes) by moderating potential damages, taking advantage of opportunities, and/or coping with the consequences.

Identifying Vulnerabilities to EPA's Mission from Climate Change

Using the best available science, EPA has begun to identify ways in which its mission, facilities, and operations are vulnerable to climate change. EPA will refine this initial effort to identify and understand areas of vulnerability by undertaking research, assessment, and monitoring activities in the coming years. EPA's *Climate Change Adaptation Plan* summarizes known vulnerabilities related to the Agency's five strategic goals:

- **Taking Action on Climate Change and Improving Air Quality.** Climate change is likely to affect air quality in a variety of ways. For example, higher air temperatures, more frequent wildfires, heavy precipitation events, and changes in atmospheric transportation patterns may lead to increased tropospheric ozone levels, changes in particulate matter exposure, worsened indoor air quality, changes to the stratospheric ozone layer, and changes in deposition of sulfur, nitrogen, and mercury in ecosystems.

> Vulnerability assessment is an ongoing process. EPA's Climate Change Adaptation Plan should be viewed as a living document that will be updated as needed to account for new knowledge, data, and scientific evidence about the impacts of climate change on EPA's mission.

- **Protecting America's Waters.** More frequent hurricanes, rising sea levels, increasing water temperatures, ocean acidification, and changing precipitation patterns may all have adverse effects on water quality protection, the operation of water management infrastructure, the quality and availability of drinking water supplies, and the restoration and protection of watersheds, wetlands, oceans, and aquatic ecosystems.
- **Cleaning Up Communities and Advancing Sustainable Development.** Flooding from sea-level rise, more intense and frequent storm surges, as well as melting permafrost could disrupt waste management networks and may also lead to releases from contaminated sites under EPA's jurisdiction. Changes in temperature and precipitation may impact the performance and efficiency of cleanups. Increased frequency and intensity of extreme weather events may affect EPA's capacity to manage debris and respond to emergencies.
- **Ensuring the Safety of Chemicals and Preventing Pollution.** Climate change may affect exposures to a wide range of chemicals due to changing use patterns and environmental conditions. For example, EPA's decisions about how pesticides are registered may be affected by the impact of climate change on pest pressure and how and where pesticides are used.

- **Enforcing Environmental Laws.** More frequent extreme weather events might divert EPA's staff and resources away from enforcement and toward disaster response and remediation.

Mainstreaming Climate Change Adaptation at EPA

Climate change is one of many factors that can influence the effectiveness of EPA's activities over time. It is essential for the Agency to account for climate change as it designs its programs, policies, and rules, in the same way that other factors such as population growth and economic development are regularly considered. EPA's *Climate Change Adaptation Plan* identifies a broad set of priority actions that EPA will take to begin addressing its vulnerabilities and mainstreaming climate change adaptation into its activities:

- **Continue to fulfill strategic measures in *FY 2011–2015 EPA Strategic Plan*:** EPA's *FY 2011-2015 Strategic Plan* established the Agency's first "Strategic Performance Measures" for integrating climate adaptation into EPA's day-to-day operations. The *Strategic Plan* committed EPA to integrating considerations about climate change science and adaptation into (1) five rulemaking processes, (2) five major grant, loan, contract or technical assistance programs, and (3) five major scientific models or decision-support tools, all by 2015.
- **Protect Agency facilities and operations:** EPA will develop and implement measures to protect its workforce, operations, and underlying infrastructure against extreme weather events such as floods.
- **Factor legal considerations into adaptation efforts:** EPA's *Climate Change Adaptation Plan* encourages managers in the Agency's programs and regions to consult with EPA's attorneys to ensure that all adaptation work is within the scope of the Agency's statutory authorities.
- **Strengthen adaptive capacity of EPA staff and partners through training:** EPA will design and implement climate change adaptation training for its staff and its partners in states, tribes, and local communities.
- **Develop decision-support tools that enable EPA staff and partners to integrate climate adaptation planning into their work:** EPA is

committed to developing decision-support tools to improve the quality and efficacy of decisions that are sensitive to changes in climate.

- **Identify cross-EPA science needs related to climate adaptation:** EPA's Office of Research and Development will coordinate an effort to identify priority research needs for the entire Agency to support the integration of adaptation planning into EPA's activities. This will produce research results that benefit multiple program areas of EPA.
- **Partner with tribes to increase adaptive capacity:** Each EPA Program and Regional Office will support the development of adaptive capacity in the tribes and identify clear steps for ongoing collaboration with tribal governments where appropriate.
- **Focus on most vulnerable people and places:** EPA will help increase the resilience of the most vulnerable people and places by improving their capacity to predict, prepare for, and avoid adverse impacts from climate change.
- **Measure and evaluate performance:** EPA will evaluate its climate change adaptation actions on an ongoing basis, in order to maintain progress toward the longterm goal of integrating climate adaptation into the Agency's programs, policies, rules, and operations.
- **Develop Program and Regional Office Implementation Plans:** EPA's National Environmental Program Offices and all of the Regional Offices have each developed their own Implementation Plans and they are being released along with this EPA Agency Plan. The Implementation Plans describe how the Offices are integrating climate adaptation into their planning and work, and how they are addressing the cross-EPA priorities identified in this Agency-wide *Adaptation Plan*. Each Implementation Plan reflects unique program or regional circumstances and goals.

MEASURING AND EVALUATING PERFORMANCE

Because of the uncertainties involved in understanding how climate change will affect EPA's programs and activities, EPA's adaptation planning will include a continuing effort of evaluation and adjustment. In the near term, EPA will evaluate its adaptation planning efforts using the three strategic performance measures from its *FY 2011-2015 Strategic Plan*. In the longer term, EPA will develop additional measures to evaluate the outcomes and effectiveness of the broader set of actions described in its *Climate Change Adaptation Plan*.

Based on these assessments, EPA will identify program areas where its climate adaptation activities have had the greatest impact, and program areas needing improvement. The lessons learned will be shared across the Agency and with EPA's partners and other stakeholders. On the basis of these lessons, EPA will make adjustments to the way adaptation is integrated into its activities.

The Importance of Program Evaluation

Because EPA's Programs and Regions will be learning by experience as they mainstream climate adaptation planning into their activities, it will be essential to evaluate their efforts in order to understand how well different approaches work and how they can be improved.

PART 1: VISION OF THE FUTURE EPA

We live in a world in which the climate is changing. Changes in climate have occurred since the formation of the planet. But humans are now influencing Earth's climate and causing it to change in unprecedented ways.

It is in this rapidly changing world that EPA is working to fulfill its mission to protect human health and the environment. Many of the outcomes EPA is working to attain (e.g., clean air, safe drinking water) are sensitive to changes in weather and climate. Until now, EPA has been able to assume that climate is relatively stable and future climate will mirror past climate. However, with climate changing more rapidly than society has experienced in the past, the past is no longer a good predictor of the future. Climate change is posing new challenges to EPA's ability to fulfill its mission.

Vision

EPA continues to fulfill its mission of protecting human health and the environment even as the climate changes.

It is essential that EPA adapt to anticipate and plan for changes in climate. It must integrate, or mainstream, considerations of climate change into its programs, policies, rules and operations to ensure they are effective under future climatic conditions. Through climate adaptation planning, EPA will continue to protect human health and the environment, but in a way that accounts for the effects of climate change.

EPA has not yet conducted a detailed quantitative assessment of the vulnerability of its mission to climate change. This Climate Change Adaptation Plan uses expert judgment, combined with information from peer-reviewed scientific literature on the impacts of climate change, to identify potential vulnerabilities. It then presents priority actions the Agency will take to begin integrating climate adaptation planning into its activities.

EPA's focus on climate adaptation is part of a larger federal effort to increase the nation's adaptive capacity and promote a healthy and prosperous nation that is resilient to a changing climate. A central element of EPA's efforts to adapt to a changing climate will be to strengthen the adaptive capacity of its own staff and its partners across the country. It will increase staff's awareness of ways that climate change may affect their ability to implement effective programs. It will empower staff to integrate climate adaptation into the work they do by providing them with the necessary data, information and tools.

Adaptive Capacity

Adaptive capacity is the ability of a human or natural system to adjust to climate change (including climate variability and extremes) by moderating potential damages, taking advantage of opportunities, and/or coping with the consequences.

EPA will also strengthen the adaptive capacity of its partners across the country in ways that are critical to attaining the Agency's mission. States, tribes, and local communities share responsibility for protecting human health and the environment, and partnerships with EPA are at the heart of the country's environmental protection system. These partnerships will be critical for efficient, effective and equitable implementation of climate adaptation strategies. EPA's Regional and Program Offices will therefore work with their partners, engage local stakeholders, and use a diversity of approaches to build adaptive capacity and encourage climate adaptation planning depending upon state, local, and tribal needs and conditions. EPA will continue to work with other federal agencies and international partners to enhance understanding of climate change risks and leverage collective knowledge about climate adaptation planning.

EPA Policy Statement on Climate-Change Adaptation

EPA issued its first *Policy Statement on Climate-Change Adaptation* in June 2011.[1] The Policy Statement recognized that climate change can pose significant challenges to EPA's ability to fulfill its mission. It called for the Agency to anticipate and plan for future changes in climate and incorporate considerations of climate change into its activities.

The Policy Statement noted that many programs throughout the Agency have already begun to anticipate and address the implications of a changing climate. These efforts laid a solid foundation on which to build climate adaptation planning into EPA's activities. Nevertheless, more needs to be done.

The *Policy Statement* called for the development and implementation of this EPA Climate Change Adaptation Plan to integrate climate adaptation into the Agency's programs, policies, rules and operations. Priority activities were also identified to be undertaken by the Program and Regional Offices, and reflected in the EPA's annual budget submissions. This Plan lays out the priority actions to begin the long-term process of integrating climate adaptation into the Agency's activities.

The *Policy Statement* also directed every EPA Program and Regional Office to develop an Implementation Plan that provides more detail on how it will meet the priorities and carry out the work called for in the agency-wide plan. The *Policy Statement* recognized that each Office is best positioned to determine how to integrate climate adaptation into its own activities, and provided each Office with the flexibility to develop its Implementation Plan in a manner consistent and compatible with its own circumstances and objectives. EPA is updating its Policy Statement to reflect the progress that has been made thus far and focus more fully on implementation activities and on helping states, tribes and communities build their capacity to adapt to climate change. The updated Policy Statement will be released in June 2014.

EPA's ongoing work is guided by principles representing EPA's core values. EPA's efforts to integrate climate adaptation into existing programs and activities will use the best available science, protect populations and locations most vulnerable to climate change and with the least ability to adapt, and use sensible analytic methods and approaches for developing adaptation strategies. Partnerships will be forged that include multiple levels of government, as well as private and nongovernmental partners throughout the country and internationally.

The *Policy Statement* acknowledges that mainstreaming adaptation planning will be an ongoing, long-term activity. The effectiveness of the Agency's adaptation activities will be monitored and evaluated to continually assess the effectiveness of actions. Lessons will be learned and shared across the Agency and with its partners at home and abroad. Likewise, EPA will learn from the experiences of its international counterparts and partners. Adjustments to the Agency's approaches and plans will be made as necessary. These adjustments will be reflected in regular updates to this agency-wide Plan.[2]

Initial Strategic Measures

EPA has established initial goals for mainstreaming climate adaptation planning into its activities. The *FY 2011-2015 EPA Strategic Plan* contains the Agency's first "strategic performance measures" for integrating climate adaptation into its day-to-day operations.[3] Explicit commitments to attain the performance measures are now included in EPA's annual budget submissions to Congress.

The strategic performance measures contained in the *FY 2011-2015 Strategic Plan* commit the Agency to integrating adaptation planning into five major rulemaking processes and five major financial assistance mechanisms by 2015, using existing authorities. They also call for the integration of adaptation planning into five major scientific models or decision-support tools used in implementing Agency environmental management programs. These *Strategic Plan* commitments represent the Agency's best and most informed judgment about the most effective mechanisms for building adaptive capacity and promoting adaptive planning within EPA and by its partners. They also provide a set of measures for monitoring the Agency's progress on adaptation planning.

EPA emphasizes the importance of evaluating activities and acting on the lessons learned. EPA will seek to identify where its climate adaptation activities might have the greatest impact on protecting human health and the environment, replicate its successes, and identify areas needing improvement. It will be an ongoing challenge to measure the direct impact of EPA's adaptation planning activities on the resilience of its programs, and on the human health and environmental outcomes it is striving to attain. The Agency will continue to explore, test and evaluate other approaches for mainstreaming adaptation planning besides those already contained in the strategic performance measures. If necessary, it will develop improved strategic measures and annual performance measures. This ongoing process of evaluation and learning is consistent with EPA's commitment to transparency and accountability.

Contribution to a Healthy and Prosperous Nation

The priority placed on mainstreaming climate adaptation within EPA complements efforts to encourage and mainstream adaptation planning across the entire federal government. Federal agencies now recognize that climate change poses challenges to their missions, operations and programs. Ensuring the capacity of federal government agencies to maintain essential services and achieve their missions in the face of climate change is critical for successful adaptation by the entire nation. Federal agencies are working together to plan for climate change using approaches that no longer assume past conditions are good indicators of the future. Although there is no single planning approach appropriate for all agencies, the use of consistent, but flexible, frameworks facilitates coordination across agencies and allows them to leverage common tools and methods.[4]

The federal government has an important and unique role in climate change adaptation, but is only one part of a broader effort that must include public and private partners throughout the country and internationally. Partnerships with states, tribes, local communities, other governments and international organizations, many of which have already begun to implement adaptation measures, are essential.

EPA's leadership and commitment to help build the nation's adaptive capacity are vital to the goal of protecting human health and the environment. Working with its partners, the Agency will help promote a healthy and prosperous nation that is resilient to a changing climate.

PART 2: KNOWN VULNERABILITIES TO EPA'S MISSION FROM CLIMATE CHANGE

2.1. Climate Change and Climate Impacts

The global climate is changing, and the impacts of this change are being felt across the United States and the world. These impacts pose new challenges to EPA as it strives to fulfill its mission of protecting human health and the environment. It is essential for the EPA to adapt if it is to reduce the vulnerability of its mission to climate change and continue fulfilling its statutory, regulatory and programmatic requirements. It is vital that the EPA anticipate and plan for future changes in climate and incorporate considerations of climate change into many of its programs, policies, rules, and operations where appropriate to ensure they remain effective under future climatic conditions.

Vulnerability

Vulnerability is the degree to which a system is susceptible to, or unable to cope with, adverse effects of climate change, including climate variability and extremes.

During the past 50 years, average temperature across the United States has risen more than 2°F, while precipitation has increased an average of about 5 percent. Some extreme weather events, such as heat waves, intense precipitation events and regional droughts, have become more frequent and intense. One of the precipitation trends in the United States is the increasing frequency and intensity of heavy downpours (the types of events that cause runoff of pollutants and pathogens into our rivers and streams, and cause combined sewer systems to overflow in our cities). This change in heavy downpours was responsible for most of the observed increase in overall precipitation during the last 50 years.[1] Also, during the past 50 years, sea level has risen up to 8 inches or more along some coastal areas of the United States, and has fallen in other locations.[2]

These trends are expected to continue, partly due to past and future emissions of heat-trapping greenhouse gases from human activities, but will occur against a background of natural variations in climate.[3] In the United States, temperatures are projected to warm substantially over the 21st century under all projections of future climate change. These changes pose risks for a wide range of human and environmental systems, including public health, the quality of the air we breathe and the water we drink, freshwater resources, the coastal environment, wildlife and ecosystems, infrastructure, economic activity, cultural resources and social wellbeing. As such, the impacts of climate change introduce vulnerabilities across the mission and goals of EPA.

Around the world all countries are expected to feel the effects of climate change, although the specific impacts will vary. The impacts, however, are expected to disproportionately affect developing countries and those already at risk.[4] Within the United States, certain parts of the population may be especially vulnerable to climate change[5]. For example, EPA recognizes that climate change may have significant impacts on subsistence resources in rural communities, Alaskan Native villages, and Indian Country. EPA's efforts to anticipate and adapt to the effects of climate change on its core mission, therefore will include helping the most vulnerable people and places reduce their exposure to climate change and improving their capacity to predict, prepare for and avoid adverse impacts.[6]

2.2. Synthesis of EPA's Vulnerabilities

The best available science directs our attention to areas where EPA's mission, facilities, and operations may be adversely affected by climate change. EPA has not yet conducted a detailed quantitative assessment of the vulnerability of its mission to climate change. This Climate Change Adaptation Plan uses expert judgment, combined with information from peer-reviewed scientific literature on the impacts of climate change, to identify potential vulnerabilities.

> The assessment of EPA's climate-related vulnerabilities is an ongoing process. This summary of known vulnerabilities should be viewed as a living document that will be updated as needed to account for new knowledge, data and scientific evidence.

This section summarizes the Agency's known mission, facility, and operational vulnerabilities. As scientific understanding increases, other vulnerabilities may join the list. This summary is organized by EPA's five strategic goals, which represent EPA's approach to its work and reflect the results it works to achieve on behalf of the American people:[7]

- Goal 1: Taking Action on Climate Change and Improving Air Quality
- Goal 2: Protecting America's Waters
- Goal 3: Cleaning Up Communities and Advancing Sustainable Development
- Goal 4: Ensuring the Safety of Chemicals and Preventing Pollution
- Goal 5: Enforcing Environmental Laws

In addition to known vulnerabilities affecting each of the five goals, EPA has begun to assess the vulnerabilities of its facilities and operations to a changing climate. EPA must ensure the safety of its personnel, the safe and continued operation of its buildings and other critical assets (e.g., vehicles), and the integrity of its grants and procurement systems. In the event of any catastrophic weather event, EPA's people, buildings and operations could be impacted. These vulnerabilities are summarized in this section. Finally, this section includes information on climate change impacts on the most vulnerable communities. For example, the Agency has a priority focus on children's environmental health and environmental justice, including minority, low-income, and indigenous populations, and these populations are discussed.

The assessment of EPA's climate change vulnerabilities is a dynamic process. The extent to which vulnerabilities have been identified and are understood varies across goals. The science of climate change will improve over time, providing greater weight of evidence to evaluate the consequences of existing and expected impacts. EPA will continue to identify new vulnerabilities and improve its understanding of known vulnerabilities as it undertakes more research, assessment, and monitoring activities, and fills in data gaps.

Examples of Data, Information, and Research Needs to Improve EPA's Assessment of its Vulnerabilities from Climate Change

Potential vulnerabilities remain difficult to assess in some areas because of limited scientific understanding of the potential impacts of climate change on some of EPA's programs. Examples of data, information, and/or research needs include:

- Characterization of local impacts to precipitation and hydrology for use in planning long-lived water infrastructure.
- Monitoring shifts in water quality and aquatic ecosystems in watersheds, and methods for incorporating such changes into

water quality programs.

- The potential impact of more intense weather events on EPA's disaster response planning efforts.
- The site-specific impacts of climate change on Brownfields, Corrective Action Facilities under the Resource Conservation and Recovery Act (RCRA), Superfund sites, RCRA Treatment, Storage and Disposal (TSD) facilities, non-hazardous solid waste facilities, and Leaking Underground Storage Tanks.
- The effect of climate change on energy efficiency programs given changes in energy demand and supply.
- The interactions between climate and the stratospheric ozone layer.
- The effects of climate change on multi-pollutant interactions in ecosystems.
- A characterization of climate-related trends in chemical use (e.g., changing patterns of pesticide use and new chemical exposures to people and the environment), and implications for the review process for new chemicals or the registration process for new pesticides.

2.2.1. Goal 1: Taking Action on Climate Change and Improving Air Quality

America's communities face health and environmental challenges from air pollution, some of which are exacerbated by the growing effects of climate change. EPA is working with its partners to protect public health and the environment with programs that address indoor and outdoor air quality, climate change, pollution prevention, energy efficiency, industrial air pollution, pollution from vehicles and engines, radon, acid rain, stratospheric ozone depletion, and radiation protection. Within this broad portfolio, several programmatic areas are vulnerable to future climate change, presenting challenges for EPA to continue to achieve its core mission.

The extent of vulnerability, however, differs across program areas and is tied to EPA's understanding of the science and projections of future climate change impacts. Some key areas of known vulnerability for EPA's air programs are:

Tropospheric ozone pollution is likely to increase in certain regions due to the effects of climate change. The relationship between temperature changes and tropospheric ozone formation is well understood. With climate change, higher temperatures and weaker air circulation in the

United States will lead to more ozone formation even with the same level of emissions of ozone forming chemicals.[8] Studies project that climate change could increase tropospheric ozone levels over broad areas of the country, especially on the highest-ozone days.[9] Climate change also has the potential to lengthen the ozone season (the months of the year when weather conditions, along with pollutants in the air, can result in the formation of ground-level ozone in particular locations around the country), and may increase individuals' vulnerability to air pollution.[10] EPA is working to reduce the number of areas in America that do not meet air quality standards. Increases in ozone due to climate change may make it more difficult to attain or maintain ozone standards. This will need to be taken into account when designing effective ozone precursor emission control programs.

Climate Change Impacts on Tropospheric Ozone Pollution

Studies project that climate change could increase tropospheric ozone levels over broad areas of the country. Climate change also has the potential to lengthen the ozone season, and may increase individuals' vulnerability to air pollution.

Photo: U.S. EPA

Sources: (1) U.S. EPA (2009). Assessment of the Impacts of Global Change on Regional U.S. Air Quality: A Synthesis of Climate Change Impacts on Ground-Level Ozone. An Interim Report of the U.S. EPA Global Change Research Program. U.S. Environmental Protection Agency, Washington, DC, EPA/600/R-07/094F. (2) K. Katsouyanni, et al., "Air Pollution and Health: A European and North American Approach (APHENA)," HEI Health Review Committee, Research Report #142 (Boston, MA: Health Effects Institute, October 2009), 5-90.

Increases in tropospheric ozone concentrations due to climate change would increase the public health burden from air pollution. The potential impacts on public health include more respiratory illnesses and increased risk of premature deaths.[11] This is a particular concern to sensitive subpopulations which are at risk for health effects from exposure to ozone. In order to better protect human health, Federal, state, tribal, and local governments will need to respond by improving the effectiveness of existing emissions control programs for ozone precursors or by implementing new control measures that will ensure attainment of the ozone National Ambient Air Quality Standards (NAAQS).

Particulate matter (PM) levels are likely to be affected through changes in the frequency or intensity of wildfires. While the impact of climate change on ambient PM levels remains somewhat uncertain, there is evidence indicating that climate change will affect PM levels through changes in the frequency or intensity of wildfires.[12] The Intergovernmental Panel on Climate Change (IPCC) has reported with very high confidence that in North America, disturbances such as wildfires are increasing and are likely to intensify in a warmer future with drier soils and longer growing seasons.[13] Forest fires are likely to increase in frequency, severity, distribution and duration in the Southeast, the Intermountain West and the West due to climate change. The potential increase in PM resulting from wildfires may increase the public health burden in affected areas, which may include sensitive subpopulations at risk for increased health effects from being exposed to PM pollution and also complicate state efforts to attain the PM NAAQS and address regional transport of air pollution.

Climate change may worsen the quality of indoor air. Climate change may worsen existing indoor environmental problems and introduce new ones as it alters the frequency or severity of adverse outdoor conditions. Some examples of potential indoor air quality impacts include:

- Heavy precipitation events may contribute to increases in indoor dampness and building deterioration, increasing occupants' exposure to mold and other biological contaminants and emissions from building materials, as well as outdoor environmental pollutants, due to breakdown of the protective building envelope.
- Temperature increases may affect the emergence, evolution and geographic ranges of pests, infectious agents and disease vectors. This may lead to shifting patterns of indoor exposure to pesticides as occupants and building owners respond to new infestations.

- Warmer average temperatures may lead to changes in occupant behavior that may create health risks. Moreover, residents may weatherize buildings to increase comfort and indoor environmental quality in addition to saving energy. Although in general, these actions should be encouraged, this may lead to a reduction in ventilation and an increase in indoor environmental pollutants unless measures are taken to preserve or improve indoor air quality.[14]

These impacts may increase public health risks, particularly to the young, the elderly, and other disproportionately impacted populations.

Climate change may alter the effects of and strategic priorities within EPA's regulatory and voluntary programs to help restore the stratospheric ozone layer. The interactions between the changing climate and ozone layer are complex. Climate change affects the ozone layer through changes in chemical transport, atmospheric composition and temperature. In turn, changes in stratospheric ozone can have implications for the weather and climate of the troposphere. Stratospheric ozone depletion and increases in global tropospheric ozone that have occurred in recent decades have differing contributions to climate change. Additionally, climate change may exacerbate the health effects of ozone layer damage at some latitudes and mitigate them at others.[15] Ozone depletion and climate change are also linked because both ozone depleting substances and most of their principal substitutes are significant greenhouse gases. While the science continues to evolve, potential climate change impacts are included in the planning and implementation of the Agency's programs to protect stratospheric ozone.

Scientific understanding related to ways that climate change may affect the interactions of sulfur, nitrogen, and mercury deposition with ecosystems is evolving. While there is limited scientific evidence on this topic, additional research is underway to better understand how patterns in the atmospheric deposition of sulfur, nitrogen, and mercury with projected changes in the climate and carbon cycle will affect ecosystem growth, species changes, surface water chemistry, and mercury methylation and bioaccumulation.[16] The potential impacts could have consequences for the effectiveness of ecosystem protection from Agency emissions reduction programs.

2.2.2. Goal 2: Protecting America' s Waters

The nation's water is the lifeblood of our communities, supporting our economy and way of life, and is the basis of all ecosystems.

EPA works with its state, local and tribal partners to protect and restore the nation's waters. Together we protect public health by reducing human exposure to contaminants in drinking water, fish and shellfish, and recreational waters. We protect and restore watersheds and aquatic ecosystems by protecting the quality of rivers, lakes, streams, wetlands and coastal and ocean waters. EPA's programs include support for drinking water, wastewater and stormwater infrastructure; setting standards for protecting water quality and regulating municipal, and industrial discharges of pollutants to waters; working to control nonpoint sources of pollution; monitoring conditions of surface water, watersheds, beaches and coastal and ocean waters; and implementing programs to preserve healthy watersheds and to restore impaired waters.

Climate change alters the hydrological cycle, changing the background conditions in which natural and man-made systems function. Changes have already been observed and are expected to continue, such as warming air and water, changes in the location and amount of rain and snow, increased intensity of rainfall and tropical storms, sea level rise, changes in ocean chemistry, and indirect effects related to energy generation and fuel production.[17]

While there is relatively high confidence in our ability to project temperature increases due to climate change, projected changes in precipitation and its effects on hydrology at the local scale are less certain. Therefore, a key challenge will be how to help local decision makers understand potential local impacts, and how to make long-term plans under a new range of uncertainty about future hydrologic conditions. Water resource managers will also need to consider the local impacts of climate change as they grapple with other challenges– including population growth, land use changes, economic constraints, and a variety of stressors to the quality and quantity of our nations waters.

Climate Change Impacts on Water

Climate change impacts include too little water in some places, too much water in other places, and degraded water quality. Some locations will be subject to all of these conditions during different times of the year. Water cycle changes are expected to continue and will affect water infrastructure, energy production and use, human health, transportation, agriculture, and ecosystems.

Photo: www.water.ky.gov
Source: USGCRP, "Global Climate Change Impacts in the U.S."(2009), Water Sector, at: http://globalchange.gov/publications/reports/scientific-assessments/us-impacts/climate-change-impacts-bysector/water-resources.

Protection of water quality, and restoration and protection of watersheds, wetlands, oceans, and aquatic ecosystems will be greatly challenged by changes in climate throughout the United States. EPA, working with its state, tribal, and local partners, is responsible for developing and implementing a portfolio of regulatory and non-regulatory programs to protect and improve water quality in the nation's watersheds and estuarine, coastal and ocean waters. As better information is developed for local decision making, changes may be needed in how EPA and our partners implement water quality programs, including Water Quality Standards, Total Maximum Daily Loads (TMDL), Effluent Guidelines, National Pollutant Discharge Elimination System (NPDES), nonpoint pollution control programs, and other watershed management programs.

The potential vulnerabilities of EPA's water quality programs to climate change include:

- Higher air and water temperatures will increase pollutant concentrations and lower dissolved oxygen levels, potentially resulting in additional water bodies not meeting water quality standards and being listed as impaired.

- Higher air and water temperatures combined with nutrient pollution will increase the incidence of Harmful Algal Blooms, threatening ecosystems and public health.
- Warmer waters and other ecological shifts will threaten aquatic habitats and aquatic species, such as cold water fisheries, with the potential for significant impacts on subsistence fishing tribes.
- Increased intensity of rainfall events and storms will cause an increase in the number of sewer overflows and wastewater bypasses, fouling streams.
- Increased intensity of rainfall events and storms will cause increased pollutant loads in runoff, and the velocity of runoff will scour and erode creek beds.
- Areas experiencing periods of less precipitation, drought, lower stream flow and limited ground water recharge will result in less water flow for dilution of permitted discharges, alterations of aquatic environments, and increased impairments. Competition will be exacerbated for limited water supplies for municipal, industrial, energy, agricultural, and ecological uses.
- Areas with increased intensity of drought or that may experience increases in events such as wildfires may see alterations in the structure and function of wetlands and watersheds.
- Sea-level rise combined with coastal development will challenge the ability of coastal wetlands to migrate.
- Ocean acidification resulting from the absorption of CO_2 will continue to stress coral reefs.
- As the nation pursues alternative strategies for producing energy and fuel, both to reduce greenhouse gases and to increase energy independence, local or regional demand for limited water supplies for energy and fuel production may increase, placing additional pressures on water quality programs.
- The ecological effects of climate change, such as shifts in aquatic species and their habitats or the quality of snowpack, are likely to affect the economic and cultural practices of tribal communities.
- Sea level rise and coastal surges increase erosion that can affect coastal zones that support aquatic species.

Climate change will have a significant impact on water infrastructure. In most of the United States, we enjoy the benefits of clean and safe water resulting from an extensive network of drinking water, wastewater and stormwater infrastructure. EPA recognizes that this infrastructure is aging and is being further taxed by the impacts of climate change. Additionally, as state, tribal and local governments face more demands for increasingly limited resources, the ability to respond to these growing infrastructure pressures becomes more complicated. Climate change will create vulnerabilities in the nation's infrastructure system in the following ways:

- Changes in rainfall patterns beyond the design capacity of drinking water, wastewater and stormwater infrastructure, or flooding due to increased intensity of storms, could overwhelm and damage infrastructure.
- Sea-level rise could affect water infrastructure, including drinking water intakes and wastewater outfalls, and could push saline water into coastal aquifers. Combined with tropical storms and associated storm surges, the integrity of coastal water infrastructure systems may be at increased risk.
- Drinking water and wastewater utilities need to take an "all hazards" approach to planning for emergencies and extreme weather events, which may be impacted by climate change. In order to support the efforts of such utilities, it is important for EPA guidance, tools, and technical support to also support this all hazards approach.
- Vulnerable and economically deprived communities may be particularly at risk, both for access to clean and safe water as well as for their ability to respond to emergencies during extreme events.

Climate change will affect the quality and availability of drinking water supplies. More than 290 million people living in the United States rely on the safety of tap water provided by public water systems that are subject to national drinking water standards. EPA ensures that these water systems are sustainable and secure by developing and revising water standards, ensuring compliance with these standards, and protecting sources of drinking water from contamination.

EPA's role in drinking water is solely to protect the quality of what Americans consume. EPA does not have a direct role in ensuring adequate water supplies. However, changes in water quantity may affect water quality. The issue of water quantity is a significant issue for many communities, and will be increasingly so especially in the west and southeast. We can expect increasing numbers of communities grappling with increased drought, reduced snow pack, and challenges to water supplies. Such communities will be faced with managing competition between municipal supplies, energy production, industrial use, agricultural use, and ecological needs, and it is likely that EPA and our partners will be called on to address water quality issues in this context. EPA has identified a number of areas where its programs designed to protect drinking water are vulnerable to climate change. These vulnerabilities include:

- Higher air and water temperatures will promote increased growth of algae and microbes, which will increase the need for drinking water treatment.
- Changes in water temperature can lead to increased risk from invasive species that can disrupt water and waste water systems.
- Increased stormwater runoff will wash sediment and other contaminants into drinking water sources, requiring additional treatment.
- Sea-level rise could increase the salinity of both surface water and ground water through saltwater intrusion, encroaching upon coastal drinking water supplies.
- Reduced annual precipitation or increased intensity and duration of drought in some regions will affect water supplies, causing drinking water providers to reassess supply plans and consider alternative pricing, allocation and water conservation options.
- Warming temperatures will cause precipitation in some areas to increasingly fall as rain rather than snow. Combined with seasonal shifts in springtime snowmelt, areas relying on snowpack to serve as a water 'reservoir' may need to develop new plans for ensuring water supplies.
- In areas with loss of snowpack or less precipitation, water demand may shift to underground aquifers or prompt development of underground storage of treated water, which will require EPA to assure the safety of such underground sources of drinking water.

2.2.3. Goal 3: Cleaning Up Communities and Advancing Sustainable Development

EPA's highest priorities under this goal are to prevent and reduce exposure to contaminants and accelerate the pace of cleanups across contaminated sites and properties, including Brownfields, Resource Conservation and Recovery Act (RCRA) Corrective Action Facilities, Superfund sites and Leaking Underground Storage Tanks.[18]

A range of major climate change stressors may affect contaminated sites, which in turn could affect how EPA addresses contamination and manages cleanups. In order to understand the potential impacts to these sites, EPA has begun to use broad screening analysis mapping to identify the sites most likely threatened by climate change impacts. EPA has a general understanding of the potential vulnerabilities at these sites. Key vulnerabilities identified by EPA include:

Flooding from more intense and frequent storms and sea-level rise, as well as melting permafrost, may lead to contaminant releases from Corrective Action sites, Superfund sites, Brownfield sites and landfills. Inundation and flooding may lead to transport of contaminants through surface soils, ground water, surface waters and/or coastal waters. Saltwater intrusion and increased ground water salinity in coastal aquifers may also increase the permeability of clay liners installed at waste sites, such as landfills, allowing contaminants to spread to nearby properties. Melting permafrost may allow contaminants to migrate and may cause land shifting and subsidence. These contaminant releases may pose an increased risk of adverse health and environmental impacts.

Increased Temperatures May Represent a Significant Risk for Waste Sites in Alaska

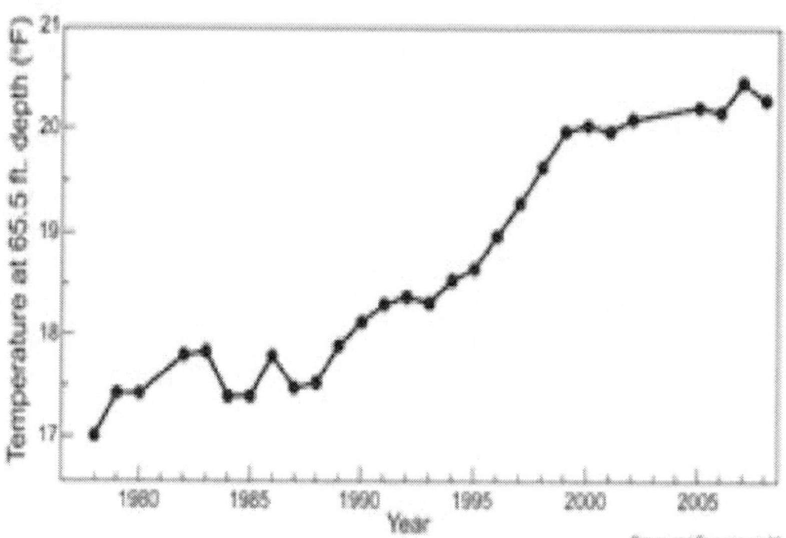

Brown and Romanovsky

Permafrost temperatures have risen throughout Alaska, with the largest increases in the northern part of the state.

Temperature increases associated with climate change may lead to the melting of permafrost – which acts as a barrier to the transport of contaminants – in northern latitudes. With increased temperatures, thawing could allow contaminants to migrate more freely to adjoining areas. Source: Thomas R. Karl, Jerry M. Melillo, and Thomas C. Peterson, eds., Global Climate Change Impacts in the United States (New York, NY: Cambridge University Press, 2009), http://downloads.globalchange.gov/usimpacts/pdfs/climate-impacts-report.pdf.

Changes in precipitation patterns and temperature may adversely affect the performance of the cleanup remedy and alter the efficacy of cleanups. To the extent that climate change leads to more prolonged droughts, water-intensive remedies may be impacted and the risk of wildfires spreading to contaminated sites may increase. Changes in precipitation may affect the rate at which vegetation grows at various sites and may affect phytoremediation and ecological revitalization efforts. The impacts may be positive or negative, depending on conditions at each site. Ground water processes may also be altered, resulting in potential adverse impacts on the performance and cost of remediation. To the extent that temperatures increase with climate change, contaminants at cleanup sites may become more volatile, increasing risks for local populations. The extent of this effect will depend on the contaminants at individual sites.

Climate change may also affect the ability of EPA's emergency management workforce to respond to natural disasters. For example, the release of hazardous substances or chemicals through flooding or storm surge creates demand for emergency response personnel and assets to monitor environmental conditions and respond. EPA has several programs to respond and minimize exposure to human populations and ecosystems, including the Oil Spill Response Program and Superfund Emergency Response. These programs provide an institutional framework to use and build upon when responding to climate change impacts. When responding to emergencies, EPA often coordinates with other Federal agencies, as well as state, tribal and local organizations. These organizations will be important partners in EPA's work responding to natural disasters of increased severity and frequency.

Flooding from more intense and frequent storms and sea-level rise may disrupt existing waste management networks Flooding from sea level rise or severe storms may disrupt the transportation system in place to handle waste. For example, flooding may disrupt the pick-up of waste in neighborhoods and business or the work performed at transfer stations. Cities with transfer stations along waterways are at particular risk. A major storm event may increase the amount of solid waste generated and lead to the release of fuel or hazardous materials. Smaller entities with hazardous materials may lack resources for emergency planning, which may increase the risk of abandoned hazardous materials during a flooding or storm event. Changes in precipitation may impact waste management practices such as composting by affecting biological processes.

2.2.4. Goal 4: Ensuring the Safety of Chemicals and Preventing Pollution

A major component of EPA's mission is ensuring the safety of chemicals. Increasingly, the chemicals used to make our products, build our homes, and support our way of life end up in the environment and in our bodies. A changing climate can affect exposures to a wide range of chemicals. Exposures may change because of changing environmental conditions or changing use patterns. EPA's efforts to reduce exposures may be affected.

EPA relies heavily on tools and models to help estimate exposures to chemicals when monitoring data are unavailable. The Agency is in the early stages of examining the ways in which its models may have to be updated to account for climate change. It has begun with a review of the potential implications of climate change for its current approaches to evaluating pesticide exposures to people and the environment.[19] These approaches are currently and will continue to be used to assess exposures to the general population, as well as children, agricultural workers and other groups who may be disproportionally affected.

EPA consulted with the Federal Insecticide, Fungicide and Rodenticide Act (FIFRA) Scientific Advisory Panel (SAP) to provide guidance on its model review and preliminary conclusions, and on sources of information that may help fill knowledge gaps.

Impact of Climate Change on Pesticide Exposure Models

Many of EPA's tools and models for examining exposure to chemicals rely on inputs that are sensitive to climate data (e.g., changing weather patterns, temperatures, stream flow rates, air currents and precipitation rates). EPA is in the early stages of examining the vulnerability of its models to climate change, beginning with a review of its pesticide exposure models.

Source: U.S. EPA, Memorandum: Transmittal of Meeting Minutes of the FIFRA Scientific Advisory Panel (SAP) held December 7, 2010, on Pesticide Exposure Modeling and Climate Change, March 3, 2011. SAP minutes, No. 2011-0.

The SAP concluded that climate change is likely to affect future decisions because of its impact on pest pressure, how and where pesticides are used, and the quantity of pesticides used. Since EPA reviews pesticide registrations every 15 years using assessment methodologies that are conservative and protective of human health and the environment, it is expected that the assessments, and decisions based on them, will remain protective. However, the SAP also concluded that weather data used in models that estimate pesticide exposure are becoming dated, and thus may not adequately reflect recent changes in climate. Some of EPA's exposure models that contain climate-related variables may have to be updated as weather patterns, temperatures, stream flow rates, air currents, precipitation rates, and other climate variables continue to change. The Agency has not yet conducted vulnerability assessments of the potential impacts of climate change on exposures of people and the environment to other types of chemicals. For example, there may be increased risk of exposure to lead and asbestos as homes, buildings, and other community infrastructure are damaged by fires, high winds, and flood events. Similarly, climate change may lead to the development of new chemicals submitted for Agency review that have uses for water purification and desalinization, wastewater treatment, antimicrobial disinfection, and disease prevention. EPA will explore the need for future assessments that evaluate potential impacts like these.

Climate Change and FIFRA

An increase in the frequency of emergency pest problems could lead to an increase in the need for emergency exemptions under the Federal Insecticide, Fungicide and Rodenticide Act (FIFRA) if currently registered pesticides are ineffective. This would allow for the use of chemicals which are not registered.

2.2.5. Goal 5: Enforcing Environmental Laws

EPA protects human health and the environment through vigorous and targeted civil and criminal enforcement and by ensuring compliance with environmental laws. Climate change may affect decisions related to the enforcement of environmental laws. For example:

- The risks posed by climate change may affect decisions by EPA about where resources should be allocated to ensure compliance with rules or regulations it believes to be priorities. These enforcement priorities may be derived from a variety of sources, ranging from the Administrator's identified goals for EPA, to program-specific guidance memoranda to assist enforcement personnel in selecting appropriate enforcement mechanisms depending on site-specific circumstances.
- A flood, hurricane or wildfire can swiftly divert the Agency's focus.[20] If climate change leads to more intense weather events and increases EPA's involvement in disaster response and remediation, then enforcement efforts (as well as efforts in other EPA programs) could be affected due to a scarcity of available staff and resources.

2.2.6. EPA' s Facilities and Operations

EPA must ensure the security of its personnel, the safe and continued operation of its buildings and other critical assets (e.g., vehicles), and the integrity of its grants and procurement systems. In the event of any catastrophic weather event, EPA's people, buildings and operations could be affected. Based on the potential for climate change to alter water supplies and increase the frequency and severity of extreme weather events, EPA has identified the following vulnerabilities to the Agency's continued safe and efficient operations:

Increased frequency and severity of extreme weather events may affect Agency facilities, personnel safety, physical security and emergency

communications. Some extreme weather events are expected to become more commonplace as the climate changes, increasing the occurrence of flooding, heat waves, lightning and high winds. An increase in these events would increase the risk to EPA's personnel in the field and EPA facilities. EPA has begun to assess these vulnerabilities and has identified the following areas of potential impact:

- Severe weather and flooding could cause damage to EPA facilities, especially in coastal areas. The Agency has already seen such damage to its Gulf Ecology Division Laboratory in Florida. Sea Level rise could also impact low lying coastal facilities and their access roads, especially when coupled with storm surges and flooding.
- Extreme weather events, including severe winds and lightning could cause damage to EPA's long-term environmental monitoring assets, particularly in coastal and flood prone areas. The Agency has already seen such damage to equipment at sites in the Clean Air Status and Trends Network and the National Atmospheric Deposition Program.
- Seasonal temperature changes and changing weather patterns can affect air quality and the general comfort of outdoor activities. Extreme heat, bad air quality or other weather conditions exacerbated by climate change may increase the health risks of EPA employees and contractors engaged in field work, such as sampling, remediation and inspections.
- Severe winds, lightning and other extreme weather events could cause power outages that disrupt EPA's security systems, outdoor lighting and emergency communication systems. Some of these systems are not linked to an uninterruptible power supply or backup generators. Outdoor lighting and security cameras are also vulnerable to direct impacts from high winds and other severe weather.

EPA's Gulf Ecology Division Laboratory

When Hurricane Ivan tore through Florida's Gulf Coast in September 2004, it served as a powerful reminder to EPA's Gulf Ecology Division Laboratory facilities that intelligent facility design in the 21st century requires the highest standards for safety and durability, as well as sustainability.

Located on Sabine Island, a 16-acre patch of land off the coast of the Florida Panhandle, the laboratory campus was especially vulnerable to the hurricane's devastating winds and rain. In the wake of the storm, six of the 40 buildings at the campus were destroyed and had to be temporarily replaced with modular structures.

In 2008, these buildings were permanently replaced with a new Computational and Geospatial Sciences Building. This facility was designed to meet the Florida Building Code, which requires stringent hurricane mitigation techniques. It has also earned the U.S. Green Building Council's Leadership in Energy and Environmental Design (LEED®) Silver 2.2 certification for New Construction. This building meets the demands of its environment in a sustainable manner.

Source: "Sustainable Facilities at EPA: Computational and Geospatial Science Building, Gulf Breeze, Florida," U.S. Environmental Protection Agency, EPA-200-F-09- 002, Washington, DC, 2009.

An increase in the number of extreme weather events could affect planning and management of emergency operations. During and after extreme weather events, EPA employees and contractors are dispatched to assess impacts to the environment and human health. The Agency also awards acquisitions and grants to support stakeholder emergency response. An increase in extreme weather events could result in the following impacts:

- An increase in the occurrence of extreme weather events may affect the availability of the Agency's personnel and resources to support the dispatch of emergency management personnel to assess environmental damage and test sites for air quality, water quality and other human health and environmental threats. At the same time, EPA personnel would increasingly be drawn away from their normal day-to-day activities to respond to extreme weather events or emergencies.
- EPA's Strategic Sustainability Performance Plan (SSPP) addresses issues related to resiliency in the Agency's suppliers, supply chain, and real property procurement under Goal 2: Sustainable Buildings and Goal 6: Sustainable Acquisition. Climate change adaptation, mitigation, and resiliency planning are incorporated into sustainable procurement efforts through the Agency's Balanced Scorecard Initiatives, and acquisition, real property and leasing decisions

incorporate sustainability through the *GreenCheck* process. Facility locations and siting decisions are driven by the National Environmental Policy Act (NEPA) process, which reviews all environmental aspects of new facilities, sites, and construction.

- Changing weather patterns and weather events may increase the demand for protective gear and appropriate vehicles and vessels to meet the demands of extreme working conditions during research, field work, and emergency management.
- EPA continues to award and manage acquisitions and grants during severe weather events; both those that are required for ongoing needs and those required for emergencies. An increase in such events could affect EPA's ability to assess contractor readiness and capabilities, process and award contracts, provide financial assistance, enter into interagency agreements and train essential personnel.

Changing water supplies may pose a risk to the quality of water used at EPA facilities. Shifts in snowpack in some regions of the country could mean a change in the disposition of water supplies and potentially compromise the quality of water available to the Agency. EPA laboratories require water to conduct experiments and meet building cooling requirements. Water shortages and quality issues could have significant impacts on the Agency's ability to manage its facilities and conduct important research, particularly in drought-prone regions.

2.3. Climate Change Impacts on the Most Vulnerable Communities

Climate change will have a disproportionate effect on particular geographic locations, communities, and demographic groups.

The impacts of climate change raise environmental justice issues. Environmental justice focuses on the health of and environmental conditions affecting minority, low-income, and indigenous populations. EPA places emphasis on these populations because they have historically been exposed to a combination of physical, chemical, biological, social, and cultural factors that have imposed greater environmental burdens on them than those imposed on the general population. Climate change is likely to exacerbate existing and introduce new environmental burdens and associated health impacts in communities dealing with environmental justice challenges across the nation.[21] EPA's *Policy Statement on Climate Change Adaptation* calls on the Agency to focus on incorporating consideration of environmental justice into the design and evaluation of adaptation strategies.

The populations most vulnerable to climate change often include, but are not limited to, the communities that are the focus of EPA's environmental justice program. Children, the elderly, the poor, the infirm, and tribal and indigenous populations are among the most vulnerable.[22] For example, children living and playing outdoors in regions with higher ozone levels resulting from increased temperature will be at higher risk for experiencing asthma symptoms and exacerbations. The elderly are more vulnerable to heat stress because they are often in poorer health, have debilitating chronic diseases and are less able to regulate their body temperature during periods of extreme heat. They may also be taking medications that increase risk for dehydration and may live alone or have fewer social contacts, which may further exacerbate their vulnerabilities.[23] Economic constraints can also place low-income households at disproportionate risk to extreme heat events due to lack of air conditioning or failure to use air-conditioning to cut down on associated energy costs.[24]

EPA has a special obligation to work consultatively with the tribes to help them as sovereign governments address their climate adaptation concerns. EPA's 1984 *Policy for the Administration of Environmental Programs on Indian Reservations* directs the Agency to work "in a manner consistent with the overall Federal position in support of Tribal 'self-government' and 'government-to-government' relations between Federal and Tribal Governments." EPA is committed to integrating environmental justice and climate adaptation into its programs, policies, rules and operations in such a way that to the extent possible, it effectively protects all demographic groups, geographic locations and communities, and natural resources that are most vulnerable to climate change. The Agency will place special emphasis on overburdened populations that are least able to help themselves, and work in partnership with them to empower them to effectively adapt to climate change.

Coastal Climate Change Impacts on Low-Income Minority Communities

Climate change will affect certain groups of people more than others, depending on where they live and their ability to cope with different climate hazards.

For example, a combination of sea level rise and land subsidence in coastal Louisiana has increased the area's vulnerability to storm surge and hurricane damage.[1,2] Hurricane Katrina, though not necessarily directly a result of climate change, provides an illustrative example of how storm surges can result in catastrophic effects for coastal communities and how social vulnerabilities can be manifested in the form greater vulnerability for lower income minority communities and unequal access to resources. Barriers for the community to avoid the risks posed by Katrina included the lack of material resources, such as cash and access to transportation, for evacuation purposes.[3,4,5] These factors contributed to disproportionate impacts on minority and low-income communities in New Orleans. For example, African Americans were overrepresented in mortality rates in all age categories compared to their proportion of the pre-Katrina population.[6] The impacts of Hurricane Katrina were devastating and highlight the environmental as well as social vulnerabilities of coastline communities.

1) Sources:USGCRP (2009). *Global Climate Change Impacts in the United States.* Karl, T.R., J.M. Melillo, and T.C. Peterson (eds.). United States Global Change Research Program. Cambridge University Press, New York, NY, USA.
2) CCSP (2008). *Impacts of Climate Change and Variability on Transportation Systems and Infrastructure: Gulf Coast Study, Phase I.* A Report by the U.S. Climate Change Science Program and the Subcommittee on Global Change Research. Savonis, M. J., V.R. Burkett, and J.R. Potter (eds.). Department of Transportation, Washington, DC, USA, 445 pp.
3) Elliott, James R. and Jeremy Pais. 2006. "Race, class, and Hurricane Katrina: Social Differences in Human Responses to Disaster," Social Science Research, 35: 295-321.
4) Fothergill, A., and L. Peek. 2004. "Poverty and disasters in the United States: A review of recent sociological findings." *Natural Hazards Journal* 32(1): 89-110.
5) Berube, A., and B. Katz. 2005 "Katrina's Window: Confronting Concentrated Poverty Across America." The Brookings Institution Special Analysis in Metropolitan Policy.
6) Sharkey, P. 2007. Survival and Death in New Orleans: An Empirical Look at the Human Impact of Katrina. Journal of Black Studies, 37: 482–501.

Climate Change Impacts on Tribal Communities

Indigenous people are among the most vulnerable communities in North America.[1] Tribes are more vulnerable to climate change impacts because of their dependence upon a specific geographic area for their livelihoods, the degree to which those geographic areas embody climate-sensitive environments, and their unique cultural, economic, or political characteristics and contexts. Also, tribes generally have fewer resources to prepare for, respond to, and recover from natural hazards, including those related to climate change.[2] The disproportionate vulnerability of tribes to climate change affects EPA's mission to protect human health and the environment in Indian country. Examples of the impacts climate change is already having on tribes include:

- The coastal Inupiat village of Shismaref Alaska is one of many coastal villages in Alaska facing relocation due to threats from flooding and erosion related to a rise in sea level and a decrease in sea ice. Sea walls have been broken and homes washed away. Residents have decided to relocate farther inland for safety, giving up their traditional fishing, sealing, and home-building sites.[3]
- Drought is perhaps the most pervasive climate-induced weather impact on tribes. Water is at the heart of many tribal cultures and the foundation of their livelihoods, economies, subsistence, and treaty rights. Water is essential to the sustainability of the fish, wildlife, and plants on which tribes rely. The recent trend toward more severe and frequent droughts, especially in the American Southwest, threatens the very underpinnings of tribal communities. The Southwest is already in the midst of a 10-15 year drought, and climate projections suggest the Southwest may transition to a more arid climate on a permanent basis over the next century and beyond.[4] In fact, climate observations indicate that this transition may have already begun.[5]
- Moose, a species important to many tribes in the Great Lakes region, are suffering the impacts of warmer weather. In a recent study of moose at the southern edge of their range in northwest Minnesota, researchers found that over the past 40 years, declines in the moose population are related to increases in mean temperature with winter and summer temperatures increasing by an average of 12°F and 4°F, respectively, over this period.

Lack of food resources and increased exposure to deer parasites associated with warmer summer temperatures appear to be the primary causes of more decline.[6]

Sources:

1) M.L. Parry, O.F. Canziani, J.P. Palutikof, P.J. van der Linden and C.E. Hanson (eds). Contribution of Working Group II to the Fourth Assessment Report of the Intergovernmental Panel on Climate Change, 2007. "Climate Change 2007: Working Group II: Impacts, Adaptation and Vulnerability." *Cambridge University Press*, Cambridge, United Kingdom and New York, NY, USA, 2007.
2) Cutter, S.L. and C. Finch. 2008. "Temporal and spatial changes in social vulnerability to natural hazards." *Proceedings of the National Academy of Science* 105(7): 2301-2306.
3) National Research Council, *Adapting to the Impacts of Climate Change*, America's Climate Choices: Panel on Adapting to the Impacts of Climate Change, The National Academies Press, Washington, DC, 2010.
4) Solomon, S., G-K Plattner, R. Knutti, and P. Friedlingstein, 2009. Irreversible climate change due to carbon dioxide emissions. *Proceedings of the National Academy of Sciences*, 106(6): 1,704-1,709. DOI: 10.1073/pnas.0812721106; Johanson, C.M., and Q. Fu, 2009: Hadley Cell Widening: Model Simulations versus Observations. *Journal of Climate*, 22:2,713–2,725.
5) Seager, R., et al., 2007. Model Projections of an Imminent Transition to a More Arid Climate in Southwestern North America. *Science*, 316: 1,181-1,184.
6) U.S. Fish and Wildlife Service. *Rising to the Urgent Challenge: Strategic Plan for Responding to Accelerating Climate Change* http://www.fws.gov/home/climatechange/pdf/CCStrategicPlan.pdf

2.4. Summary of Key Areas of Known Vulnerability

The current list of known vulnerabilities of EPA's programs to climate change are summarized in the table at the end of this report. The vulnerabilities listed in the table help to guide the Agency in identifying areas to focus its adaptation planning efforts. For several of the vulnerabilities, current scientific understanding is that the climate impact is likely or very likely to occur and EPA's best judgment is that there is a high likelihood the program will be affected. The Agency, as part of its efforts to mainstream adaptation into its programs (addressed in Part 3 of this document), will conduct a more comprehensive vulnerability assessment to determine which programs and areas are most suitable to initiate action. This qualitative assessment has been done at a national level. It identifies vulnerabilities to entire programs within EPA to help focus the Agency's climate adaptation efforts. However, there is a "regional texture" to the impacts of climate change. The severity and importance of known vulnerabilities will vary across regions. EPA's Program and Regional Climate Change Adaptation Implementation Plans capture the regional differences and identify the vulnerabilities of greatest importance, including identifying the most vulnerable people and places within these programs. The Implementation Plans describe how climate change adaptation will be integrated into Program and Regional planning and work in a manner consistent and compatible with their own circumstances and objectives.

PART 3: MAINSTREAMING CLIMATE CHANGE ADAPTATION IN EPA

3.1. Building Adaptive Capacity

EPA's *Policy Statement on Climate-Change Adaptation* acknowledges the importance of adapting to climate change if the Agency is to continue fulfilling its statutory, regulatory and programmatic requirements. It is vital that the EPA anticipate and plan for future changes in climate and incorporate considerations of climate change into many of its programs, policies, rules and operations to ensure they remain effective under future climatic conditions.[1]

Climate change is one of many factors that can influence the effectiveness of EPA's activities over time. It is essential the Agency account for climate change as it designs its programs, policies and rules, in the same way other factors such as population growth and economic development are regularly considered. EPA will integrate, or mainstream, climate change adaptation by strengthening the adaptive capacity of its own staff. EPA will empower them to account for climate change in the normal course of doing business. It will increase staff's awareness of ways that climate change may affect their ability to implement effective programs, and provide them with the necessary data, information and tools to integrate climate adaptation into the work they do.

EPA will work to strengthen partners' adaptive capacities. The Agency will work with its state, tribal and local partners and will strive to ensure that to the greatest extent possible, their human health and environmental protection programs are resilient to climate change. This is particularly important since the Agency authorizes many states and tribes to implement various environmental programs. Also, EPA's efforts to help communities become more environmentally and economically sustainable could be affected by the impacts of climate change. EPA will work with its partners in states, tribes and communities to look for opportunities to create co-benefits from implementation of adaptation plans. For instance, where communities take steps to improve storm water system resilience by reducing combined sewer overflows, these actions may also reduce energy use for water. EPA will need to ensure that its tools, research, and technical assistance evolve to help communities and other entities take projected climate changes into account as they plan development. International partnerships and collaboration on adaptation will also be important to address the transboundary impacts of climate change.

The Agency's investment in building adaptive capacity is an ongoing effort. It will require a sustained, long-term commitment to empower EPA's workforce and partners by providing the information and tools necessary to account for climate change.

3.2. Guiding Principles for Adaptation at EPA

EPA has adopted a set of principles to guide all of its efforts to integrate climate adaptation into its programs, policies and rules. The principles affirm EPA's approach of integrating climate adaptation into existing programs and activities to ensure their effectiveness as the climate changes. They uphold EPA's core values of using the best available science, protecting populations and locations most vulnerable to climate change, and using sensible analytic methods and approaches for developing and implementing adaptation strategies. EPA has an important and unique role in climate adaptation, but is only one partner in a broader effort that must include multiple levels of government, as well as private, nongovernmental, and international partners. The principles call for ongoing evaluation of the effectiveness of climate change adaptation approaches, recognizing that the Agency will continue to learn how to adapt effectively over time.

Guiding Principles for Adaptation

- Adopt integrated approaches: Adaptation should be incorporated into core policies, planning, practices and programs whenever possible.
- Prioritize the most vulnerable: Adaptation plans should prioritize helping people, places and infrastructure that are most vulnerable to climate impacts and be designed and implemented with meaningful involvement from all parts of society.
- Use best-available science: Adaptation should be grounded in the best-available scientific understanding of climate change risks, impacts and vulnerabilities.
- Build strong partnerships: Adaptation requires coordination across multiple sectors and scales and should build on the existing efforts and knowledge of a wide range of public and private stakeholders.
- Apply risk-management methods and tools: Adaptation planning should incorporate risk-management methods and tools to help identify, assess and prioritize options to reduce vulnerability to potential environmental, social and economic implications of climate change.
- Apply ecosystem-based approaches: Adaptation should, where relevant, take into account strategies to increase ecosystem resilience and protect critical ecosystem services on which humans depend to reduce vulnerability of human and natural systems to climate change.
- Maximize mutual benefits: Adaptation should, where possible, use strategies that complement or directly support other related climate or environmental initiatives, such as efforts to improve disaster preparedness, promote sustainable resource management, and reduce greenhouse gas emissions including the development of cost-effective technologies.
- Continuously evaluate performance: Adaptation plans should include measureable goals and performance metrics to continuously assess whether adaptive actions are achieving desired outcomes.

Source: The White House Council on Environmental Quality, "Progress Report of the Interagency Climate Change Adaptation Task Force: Recommended

Actions in Support of a National Climate Change Adaptation Strategy," October 5, 2010, http://www.whitehouse.gov/sites/default/files/microsites /ceq/Interagency-Climate-ChangeAdaptation-Progress-Report.pdf.)

3.3. Agency-Wide Priorities

EPA has identified priority actions it will take to begin integrating climate change adaptation into its programs, policies, rules and operations. These priorities represent EPA's commitment to address the known vulnerabilities of its mission to climate change, and to continue to identify other vulnerabilities its programs may have to climate change.

Agency-wide Priorities

- Fulfill Strategic Measures in FY 2011–2015 EPA Strategic Plan
- Protect Agency facilities and operations
- Factor legal considerations into adaptation efforts
- Strengthen adaptive capacity of EPA staff and partners through training
- Develop decision-support tools that enable EPA staff and partners to integrate climate adaptation planning into their work
- Identify cross-EPA science needs related to climate adaptation
- Partner with tribes to increase adaptive capacity
- Focus on most vulnerable people and places
- Measure and evaluate performance
- Develop and Implement Program and Regional Office Implementation Plans

3.3.1. Priority: Fulfill Strategic Measures in FY 2011-2015 EPA Strategic Plan

EPA expects that its understanding of how to integrate climate adaptation into its programs, policies, rules and operations will improve over time. The FY 2011–2015 EPA *Strategic Plan* identified three initial mechanisms through which the Agency will begin mainstreaming climate adaptation by 2015:[2]

Strategic Measures in FY 2011-2015 EPA Strategic Plan

By 2015:

1) Integrate climate change adaptation into rulemaking processes.
2) Integrate climate adaptation into financial mechanisms.
3) Develop decision-support tools.

1. Integration of climate change adaptation into rulemaking processes: EPA will integrate climate change trend and scenario information into five rulemaking processes to further EPA's mission, consistent with existing authorities. The integration of climate adaptation into Agency rulemaking processes will help ensure the rules are effective as the climate changes.[3] A variety of "entry points" can be considered, including the development of the rule itself; related policy and guidance development; and post-rule permitting, monitoring and enforcement.

EPA will enhance the ability of rule makers to address the implications of climate change through updates to the Action Development Process (ADP). This process was developed by EPA to guide the Agency's rulemaking activities from the start of the rulemaking process through the analysis of regulatory options to the final publication of a regulation. EPA will integrate climate adaptation into the ADP by:

- **Identifying process points where climate change adaptation considerations need to be identified, analyzed and discussed.** The rulemaking process[4] includes opportunities to discuss climate change adaptation considerations, both internally and with stakeholders. An Analytic Blueprint spells out a workgroup's plan for data collection and analyses to support development of a specific action. The development of an Analytic Blueprint provides an early opportunity to articulate any climate change adaptation issues that need analysis.
- **Developing guidance documents and training rule writers to understand the implications of climate change impacts.** EPA has guidance on addressing children's health and environmental justice for all of its rulemakings. EPA will develop a similar guide for climate change adaptation, and provide training. This will improve the regulatory work groups' understanding of climate change adaptation and how to consider it in rulemaking, when appropriate.

Progress is already being made to fulfill the Strategic Measures

EPA is already making progress to fulfill the three Strategic Measures. For example, the Agency issued guidance in October 2011 encouraging all Offices to include climate adaptation evaluation criteria into announcements of competitive funding opportunities The guidance is relevant to announcements in which the outcomes to be supported by the awards are sensitive to changes in climate (e.g., ability to attain air quality standards; effectiveness of water infrastructure), or the projects being solicited would be more effective if they addressed climate change adaptation issues (e.g., development of models and tools to support decision making). EPA is also making progress on integrating climate adaptation into its categorical funding mechanisms such as the Clean Water and Safe Drinking Water State Revolving Funds in the Water Program and Brownfields Restoration Grants in the Office of Solid Waste and Emergency Response.

EPA is also already developing tools to support climate adaptation planning. For example, the BASINS tool that is designed for use by regional, state, and local agencies in performing watershed and water quality-based studies now includes a Climate Assessment Tool (CAT). CAT provides a capability for understanding how water resources could be affected by a range of potential changes in climate, and the possible effectiveness of management practices for increasing resilience of water resources to climate change.

- **Tracking and monitoring rulemakings where climate change adaptation may be an important consideration.** To help identify rulemakings where climate change adaptation may be relevant, EPA will use its internal regulatory tracking databases to manage and report on regulatory development. This will allow EPA workgroups and stakeholders to plan and allow for integration of climate change considerations when identified in the critical process points described above.

2. Integration of Climate Change into financial assistance mechanisms: EPA will integrate considerations of climate change impacts and adaptive measures into five major grant, loan, contract or technical assistance programs. This will further EPA's mission, consistent with existing authorities. The integration of "climate adaptation criteria" into financial mechanisms will encourage recipients to account for climate change.

Although this Strategic Measure is limited to a goal of five major financial assistance mechanisms by 2015, the ultimate goal is to integrate climate adaptation into all financial assistance agreements, where appropriate.

3. Development and use of decision-support tools: EPA will integrate climate change trend and scenario information into five major scientific models or decision-support tools used to implement Agency environmental management programs. The development of decision-support tools will help build the adaptive capacity of the Agency's workforce and its partners.

These three mechanisms represent different pathways through which the Agency can integrate climate change into ongoing programs and priorities, to attain desired environmental and human health outcomes and sustain them as the climate changes. EPA will continue to explore more pathways through which the resilience of the Agency's mission can be enhanced.

3.3.2. Priority: Protect Agency Facilities and Operations

EPA is committed to the safety of its personnel, the integrity of its buildings, and the efficiency of its operations, but the increasing frequency and severity of extreme weather events poses risks to meeting these objectives. Climate change could disrupt the operation of the Agency's programs, compromise the safety of its staff, or affect the integrity of its physical infrastructure. Adaptation planning to protect EPA's workforce, operations and underlying infrastructure is crucial.

EPA will develop and implement measures to protect its workforce and increase the resilience of its facilities and operations to climate change. For example, where possible, EPA will enhance the resilience of existing facilities in coastal areas to protect them from severe weather, flood damage, and sea level rise. The Agency will also work with other government agencies, particularly the General Services Administration, to account for climate change in the design and construction of new facilities, or when new buildings are leased.

3.3.3. Priority: Factor Legal Considerations into Adaptation Efforts

As a general matter, the Agency's broad mandates to protect human health and the environment afford a large reservoir of legal authority to support EPA adaptation work. However, specific legal questions that may arise in the course of adaptation issues cannot be answered in the abstract.

As a federal agency, EPA derives its authority to act from the U.S. Constitution and the laws passed by Congress. The Agency is committed to ensuring that its actions are constitutional, authorized by statute, consistent with Congress's vision and intent, and otherwise legally supported. The 2011 EPA *Policy Statement on Climate-Change Adaptation* called on the Agency to "identify for the Office of General Counsel areas where legal analysis is needed to carry out agency actions called for in this policy statement." Because the legality of its actions is such a high priority for EPA, program managers and staff are encouraged to freely and frequently consult with the appropriate attorneys in the Office of General Counsel (OGC), Offices of Regional Counsel (ORC), and the Office of Enforcement and Compliance Assurance (OECA) as they conduct their adaptation work.

Important variation exists among the statutes EPA administers, as well as the regulatory programs EPA designs, implements, and enforces under those laws. Some of these laws, like the Clean Air Act and the Clean Water Act, give EPA regulatory powers, such as the authority to write regulations, set standards, issue permits, ensure compliance, and authorize state and tribal environmental programs. Other laws govern EPA actions in a variety of areas essential to its mission, such as research and development, budget and personnel management, contracts, and the award of financial assistance. Still other laws impose obligations on EPA, such as responsibilities to evaluate the effect of its activities on state and local governments, overburdened communities, small businesses, and endangered species, among others.

Each of these laws, whether granting EPA authority or imposing an obligation, may deserve special attention and analysis in resolving legal questions related to adaptation work. For example, EPA may need to determine the extent of its authority to incorporate adaptation measures into the terms and conditions of financial assistance mechanisms; evaluate the legal basis for considering climate change impacts in setting standards or issuing permits under the Clean Air Act and Clean Water Act; or review the adequacy of its emergency response authorities in the context of more frequent natural disasters. These examples are merely illustrative of the diversity of ways in which legal issues may arise for EPA as it mainstreams climate adaptation.

Important variation also exists in the level of scientific understanding of climate change impacts and the sensitivity of EPA programs to those impacts. These variations inform the analysis of EPA's legal authority and responsibilities. For instance, under the Administrative Procedure Act, federal agencies like EPA have a basic obligation to act transparently and rationally. This is generally demonstrated through an administrative record that documents the analysis and reasoning leading to a final decision and responds appropriately to concerns raised by interested parties and the public, using the information available to the Agency at the time of its decision. The relative weight climate change considerations should be given in evaluating options for EPA action may depend on factors such as the time and geographic scale of the potentially relevant climate impacts compared to the temporal and spatial scale of the proposed EPA action; the strength of the scientific understanding of the climate impacts; and the environmental and economic consequences estimated to result from including or choosing not to include climate change adaptation measures or considerations in the EPA action.

Considerations such as these are by definition case specific. Over time, however, EPA anticipates that more detailed policy principles and legal precedents will emerge to further guide and inform EPA's adaptation efforts.

3.3.4. Priority: Strengthen Adaptive Capacity of EPA Staff and Partners Through Training

An organization with adaptive capacity has the ability to craft and adopt new means to achieve its goals as circumstances change. EPA needs its personnel and partners in states, tribes, and local communities to have adaptive capacity if it is to achieve its mission in the midst of climate change. EPA will build adaptive capacity through ongoing education and training. Equipped with an understanding of expected climate-related changes and adaptation approaches, and provided with and trained on how to use new decision-support tools, EPA and its partners will be able to incorporate climate change adaptation into their plans and decisions. EPA's training, education and outreach programs that are focused on climate adaptation will evolve over time. As an initial step, EPA will design and implement a training program for its staff and its partners focused on topics relevant to EPA's mission. One goal is to increase awareness about the importance of climate change adaptation, and to encourage all EPA staff and partners to consider the changing climate in the normal course of business. A second goal of EPA's training will be to expose its staff and partners to specific approaches and tools for integrating climate adaptation into decision-making processes.

3.3.5. Priority: Develop Decision-Support Tools that Enable EPA Staff and Partners to Integrate Climate Adaptation Planning into their Work

The fact that the climate is no longer relatively stable, but will continue to change in new ways, presents a major challenge for decision makers working to protect human health and the environment. Many standard practices may no longer be effective unless they account for climate change. For example, standard methods used for estimating the probability and expected frequency of floods for flood plain mapping, designing infrastructure systems, and estimating runoff of pollutants and sediments into rivers and streams are based on the assumption of climate stationarity.[5] The end of climate stationarity means that EPA and its partners need to alter their standard practices and decision routines to account for a continuously changing climate.

The assumption of stationarity

Until now, EPA (for those programs not explicitly focused on climate change) has been able to assume stationarity of climate; that is, climate is relatively stable and future climate will mirror past climate. But the past is no longer a good predictor of the future.

The development of decision-support tools plays a central role in EPA's overall efforts to adapt to climate change. Following the recommendations of the National Research Council, EPA is committed to developing decision-support tools to improve the quality and efficacy of decisions related to outcomes that are sensitive to changes in climate.[6] These tools will empower staff to consider climate, as well as changes in social and economic conditions that are influenced by climate change. They will enable staff to integrate climate adaptation planning into their work and decision-making processes. Priority will be given to the development of tools that would benefit end-users in multiple areas of EPA.

Decision Support Tools

"The effectiveness of any decision support tool depends on whether it provides information that is relevant to decision makers. Tools need to be useful at space and time scales that are meaningful and relevant for specific decisions and decision makers, and they also need to be based on up-to-date and reliable information"

Source: National Research Council, *Informing an Effective Response to Climate Change*, America's Climate Choices: Panel on Informing Effective Decisions and Actions Related to Climate Change, The National Academies Press, Washington, DC, 2010.

EPA will also support capacity-building for state, tribal, local, and international partners by working with them to develop and use effective decision-support tools. EPA will coordinate with other Federal agencies on developing decision-support tools with partners, when appropriate.

3.3.6. Priority: Identify Cross-EPA Science Needs Related to Climate Adaptation

Implementing effective strategies to adapt to the changing climate requires that decisions be grounded in the best available science on climate change risks, impacts and vulnerabilities, and adaptive management practices. Throughout EPA, there is a growing need for up-to-date information on the existing models, tools, data and information relevant to climate change adaptation.

EPA has made great progress in climate-related research and with the development of models and tools. However, the complex interactions of climate change impacts mean that uncertainties and data gaps persist and that multiple Agency stakeholders have a role to play in developing a research agenda. In order to identify the most pressing science needs for improved adaptation decision making, *priority* research needs related to climate change adaptation will be identified and periodically updated for the *entire* Agency through a coordinated approach. This approach is designed to produce research results that benefit end-users in multiple areas of EPA.

EPA will advance a rigorous basic and applied science program that will inform, enable and deliver innovative and sustainable solutions to environmental problems in a changing climate. The EPA Office of Research and Development (ORD) has the primary responsibility of coordinating with the Program and Regional Offices to identify the priority science needs of the Agency and its partners. This coordination is essential since some of the priority science needs will be met by ORD's research program, and some by scientists in Program and Regional Offices. In cases where other agencies could produce the scientific information needed, ORD will play a major role representing EPA's needs to other federal agencies. For example, ORD is EPA's primary representative to the U.S. Global Change Research Program (USGCRP), which coordinates and integrates climate change research across 13 federal departments and agencies.

Available data, tools, and information will be shared across the Agency and with its partners to avoid redundancy given the Agency's limited resources. To facilitate the ongoing sharing of information, EPA will establish a central repository of data (and associated metadata), models, tools and information related to climate adaptation that are produced by the Agency. The repository will also include information ("lessons learned") about methods for mainstreaming climate adaptation that have been used by particular EPA Offices that may be applicable to other users across the Agency. Where relevant to the adaptation efforts of EPA and its partners, the repository will be linked to other databases and repositories of information within EPA, as well as those produced by other federal agencies and non-federal entities.

3.3.7. Priority: Partner with Tribes to Increase Adaptive Capacity

EPA is committed to strengthening its partnerships with tribes on their priorities related to climate change adaptation. A unique government-to-government relationship exists between the U.S. Government and the 565 federally recognized tribes. EPA gives special consideration to tribes in developing policies that may affect their interests. EPA recognizes that tribes will likely be disproportionately vulnerable to climate change. This disproportionate vulnerability is partly due to their dependence on specific geographic areas for their livelihood; unique cultural, economic and political characteristics; and limited resources to prepare for, respond to and recover from climate-related hazards (*i.e.*, limited adaptive capacity).[7]

Each Program and Regional Office Implementation Plan will support the development of adaptive capacity in the tribes and identify clear steps for ongoing collaboration with tribal governments where appropriate. These efforts will include increasing tribal capacity to identify vulnerabilities in order to adapt to a changing climate. EPA will work with tribes to support the effectiveness of national climate change adaptation programs in Indian country. The Agency will support the development of climate science to meet priority research needs and decision-support tools useful to the tribes. EPA will also work with the tribes to identify and support the use of climate change relevant traditional ecological knowledge (TEK) in decision making. EPA recognizes that TEK, as an expression of key information that links historical, cultural and local ecological conditions, may help tribes choose how they adapt to climate change while also protecting resources and resource uses important to their culture and livelihood. These efforts will leverage existing EPA partnerships with the tribes and tribal networks.

On a national level, EPA will work with other Federal agencies to collectively support tribes as they assess their vulnerabilities to climate change and plan and implement adaptation actions. Regional Offices will seek opportunities to work together with other Federal agencies' regional offices to provide strong support to tribes on their particular climate change challenges.

3.3.8. Priority: Focus on Most Vulnerable People and Places

The Agency places special emphasis on, and works in partnership with, overburdened populations. As discussed in Part 2 of this report, certain parts of the population, such as children, the elderly, the poor, Tribes and indigenous people can be especially vulnerable to the impacts of climate change. This may be due to susceptibility to health impacts of environmental contaminants, economic status, health status, education or access to information. Also, certain communities will be particularly vulnerable, such as those that are in low-lying coastal areas or subsistence fishing tribal communities. EPA is also concerned about the potential impacts on human health and the environment in rural communities, including agricultural communities. In these communities, some climate change effects such as long-term drought or severe storms have the potential to cause severe effects on local economies. EPA's efforts to anticipate and adapt to the effects of climate change on its core mission, therefore, will include helping the most vulnerable people and places reduce their exposure to climate change and improving their capacity to predict, prepare for and avoid adverse impacts. For example, Program and Regional implementation plans will include a focus on understanding the environmental-justice implications of climate change impacts, identifying populations and communities vulnerable to climate change and with limited ability to adapt, and incorporating consideration of environmental justice issues into the design and evaluation of adaptation strategies.

The Agency will make special efforts to connect with populations that have been historically underrepresented in decision-making in order to support the development of adaptation plans that are culturally sensitive and that improve their capacity to predict, prepare for, and avoid climate change impacts. The Agency will also continue to focus on life stages vulnerable to climate change.[8] The development of effective adaptation plans and strategies will also be improved by examining the interaction of multiple stressors, including climate change, on communities and populations.

3.3.9. Priority: Measure and Evaluate Performance

Evaluation is a systematic way to learn from experiences. In its *Strategic Plan*, EPA emphasizes the importance of evaluating activities and acting on the lessons learned. Through systematic evaluation, the Agency can identify where activities have the greatest impact on protecting human health and the environment; provide the roadmap needed to replicate successes; and conversely, identify areas needing improvement.

EPA will evaluate its climate change adaptation actions on an ongoing basis to assess the Agency's progress toward attaining the desired long-term outcome of mainstreaming climate change adaptation into the Agency's programs, policies, rules and operations. Based on lessons learned about the most effective climate change adaptation strategies, EPA can make adjustments to the way adaptation is integrated into its activities.

EPA's commitment to measuring and evaluating the progress it is making to integrate climate adaptation into its programs, policies, rules, and operations is discussed in greater detail in Part 4.

3.3.10. Priority: Develop and Implement Program and Regional Office Implementation Plans

EPA includes National Environmental Program Offices and National Support Offices headquartered in Washington, D.C., and 10 Regional Offices around the country. The 2011 EPA *Policy Statement on Climate-Change Adaptation* directed the development of programmatic Implementation Plans. EPA National Environmental Program Offices[9] and Regional Offices developed their own Implementation Plans providing more detail on how they will integrate climate adaptation into their planning and work, and help address the cross-EPA priorities identified in this agency-wide plan. The National Environmental Program and Regional Offices had flexibility to develop their Implementation Plans in a manner consistent with their own circumstances and objectives. Support Offices will assist with the implementation of the Agency-wide Plan and National Environmental Program and Regional Office Plans as they do for other initiatives. The ultimate goal of the Implementation Plans is to focus on the core missions and priorities of the Program and Regional Offices, and to ensure that their programs and operations are resilient and effective in a changing climate.

While flexibility was essential to produce Implementation Plans that are effective and responsive in meeting each specific organization's needs, Environmental Program and Regional Offices cooperated as they developed their respective plans, shared experiences and lessons learned, and coordinated on issues that cut across Agency programs and regions. The EPA National Water Program developed a second iteration of its National Water Program Strategy: Response to Climate Change. The 2008 Strategy and 2012 revision provided opportunities for lessons learned and examples of how other EPA programs could approach development of their plans.[10]

The Cross-EPA Work Group on Climate Adaptation Planning will oversee the development of the Implementation Plans and identify required interim products (e.g., comprehensive vulnerability assessments) that will facilitate the development of the Plans.

Common Areas of Focus for Implementation Plans
1) Vulnerability assessments 2) Priority actions on climate adaptation 3) Agency-wide Strategic Measures on climate adaptation 4) Legal and enforcement issues 5) Training and outreach 6) Partnerships with tribes 7) Vulnerable populations and places 8) Evaluation and cross-Office pilot projects

Annual highlights of progress from each Environmental Program and Region will be included in updates to the agency-wide Climate Change Adaptation Plan that will be submitted on an annual basis with the Agency's Sustainability Plan. The Implementation Plans will also be used to provide input to the Agency's annual planning and budgeting process, where appropriate.

In order to promote consistency, the Implementation Plans for all Environmental Program and Regional Offices have several common areas of focus, as outlined in the table below. There is diversity among the plans and some offices may have a broader scope in some areas than others. For instance, Regional Offices focus their efforts on particular geographic locations more than National Program Offices.

1) **Vulnerability assessments:** Each Implementation Plan (Office or Region) contains an initial assessment of the implications of climate change for the organization's priorities and objectives. This assessment builds on the work presented in Part 2 of this document. Program and Regional Offices are at different stages of understanding and addressing the ways climate change may affect their respective missions. Some Implementation Plans therefore have more detailed information on vulnerabilities than others.
2) **Priority actions for climate adaptation:** Each Implementation Plan describes the organization-specific priorities related to climate change adaptation. At the core of each Implementation Plans a description of the activities that the Program or Regional Office will pursue over

time to integrate climate change adaptation into its programs, policies and operations. The Plans describe how these activities address both organization-specific priorities and the cross-Agency priorities. In addition, Implementation Plans describe how Program and Regional Offices will work together on actions that are most effectively accomplished by more than one Office or Region.

For each action, the Implementation Plan identifies the organization's key partners at the international, federal, state, tribal, local, public and private sector levels, including state, tribal and local co-regulators. Attention will be given to engaging those partners who have been historically under-represented.

Activities include both short- and long-term actions. Short-term activities include actions that are readily achievable, such as specific training needed to begin building adaptive capacity. Short-term activities also focus on areas where the organization has relative certainty about climate impacts, and therefore feels that action cannot be delayed. The more immediate actions enable the organization to learn what works. Armed with the lessons learned, the organization can move forward with insights and information as it tackles additional issues. Longer-term activities will focus more broadly on building resilient, healthy communities that have the knowledge and tools needed to inform decisions.

3) **Agency-wide Strategic Measures on climate adaptation:** At a minimum, the Implementation Plans consider activities that address the three existing Agency-wide Strategic Measures (discussed in Section 3.3.1 and in Part 4), to the extent they are applicable to the organization.
4) **Legal and enforcement issues:** Specific legal questions may have arisen as each Program and Regional Office integrates adaptation planning into its programs, policies, and rules. Each Office describes how its program managers and staff will consult with the appropriate attorneys in the Office of General Counsel (OGC), Offices of Regional Counsel (ORC), and the Office of Enforcement and Compliance Assurance (OECA), as they conduct their adaptation work.
5) **Training and outreach:** All Program and Regional Offices will benefit from Agency-wide training activities that they will work together to develop and implement under the auspices of the Cross-EPA Work Group on Climate Change Adaptation. Each

Implementation Plan describes the ways in which the organization will use the Agency-wide training resources to educate its staff. Each Implementation Plan also indicates how the organization will then, over time, integrate climate adaptation where appropriate into *existing* Office-specific training programs used by its workforce and external partners. Regional Offices, working in coordination with HQ Program Offices as needed, may also choose to take the lead on cross-media training and awareness-building among states and other external partners.

6) **Partnerships with tribes:** Each Program and Regional Office Implementation Plan includes actions to address the tribes' adaptation issues relevant to the Office. The Implementation Plans identify how the Office will work collaboratively with tribes to increase the adaptive capacity of the tribes. This partnership will help ensure that priority tribal adaptation needs are addressed, and efforts to build adaptive capacity within tribes are effective. The Office of International and Tribal Affairs (OITA) will support the efforts of all Offices to consult and partner with the tribes to develop and implement the actions. Also, OITA will help coordinate the interactions of EPA Offices with tribes to promote unified EPA consultations with individual tribes.
7) **Vulnerable populations and places.** Each implementation plan describes how the Program or Regional office will identify vulnerable populations and places to climate change. The process of conducting vulnerability assessments and determining priority actions for climate adaptation should consider how each Program and Regional Office can help vulnerable populations and places reduce their exposure to climate change and how to improve their capacity to predict, prepare for, and avoid adverse impacts. The plans are a useful tool to account for the regional variability in how climate change will impact people and places already overburdened by environmental pollution or other stressors.
8) **Evaluation and cross-organization pilot projects:** Each Implementation Plan includes a process for measuring and evaluating the effectiveness over time. Program and Regional Offices are encouraged to partner with each other to conduct pilot projects that test climate adaptation approaches that are broadly applicable. Relevant Implementation Plans describe these joint efforts, as well as Office-specific pilot projects. The goal is to learn what approaches

work and why. The Implementation Plans will periodically be adjusted to improve the organization's efforts to integrate climate adaptation into its activities.

3.4. Importance of Partnerships

EPA believes strong partnerships are critical to fulfilling its mission of protecting human health and the environment. As stated in the EPA Strategic Plan, successful partnerships make the most effective use of partners' respective bodies of knowledge, resources and talents. Partnerships are keys to effective integration of climate change adaptation considerations into the protection of human health and the environment. In general, EPA will focus adaptation work on existing geographic-based partnerships with the goal of increasing effectiveness in climate change adaptation efforts.

Climate Change Collaboration - One potential model of place-based collaboration is C3 (the Climate Change Collaboration) in EPA Region 10 (Seattle, Washington). C3 includes 13 agencies focused on climate change that meet monthly (http://www.c3.gov/). The group was formed through a bottom-up approach when representatives from several agencies realized they were working (redundantly) on the same topics. Through its participation in C3, Region 10 has gained better access to information and technical assistance to support adaptation efforts at local and regional levels; it was also able to learn from the adaptation efforts of states and communities within and outside the region. Similar efforts beyond Region 10 include those in New England, the Southeast, the Great Lakes, and the Gulf Coast.

States, tribes, and local communities share responsibility for protecting human health and the environment, and partnerships with EPA are at the heart of the country's environmental protection system. These partnerships will be critical for efficient, effective and equitable implementation of climate adaptation strategies. EPA's Regional and Program Offices will therefore work with their partners, engage local stakeholders, and use a diversity of approaches to form the development of adaptive capacity and encourage climate adaptation planning depending upon state, tribal, and local needs and conditions.

EPA will continue to build and maintain strong partnerships with other federal agencies. For example, EPA will continue to actively participate in the interagency Council on Climate Preparedness and Resilience established under EO 13653 and related working groups. EPA is part of the Federal Agency Climate Change Adaptation Community of Practice, a forum that allows agency staff working on adaptation to share knowledge and experience on adaptation planning, implementation and evaluation. The Program and Regional Office Implementation Plans identify specific areas of collaboration with other federal agencies, such as in delivering support to tribes.

Finally, climate change impacts do not stop at our borders, but instead can pose risks globally. EPA is committed to working with our international partners to share expertise, practical experiences, information and data to address adaptation issues.

Cross-cutting national strategies relevant to adaptation planning at EPA

- President's Climate Action Plan
- Executive Order 13653 - Preparing the United States for the Impacts of Climate Change
- National Action Plan: Priorities for Managing Freshwater Resources in a Changing Climate
- Draft National Fish, Wildlife and Plants Climate Adaptation Strategy
- Draft National Ocean Policy Implementation Plan

For more information: http://www.whitehouse.gov/administration/eop/ceq /initiatives/adaptation)

PART 4: MEASURING AND EVALUATING PERFORMANCE

4.1. Existing Strategic Performance Measures

The *FY 2011-2015 EPA Strategic Plan* outlines the Agency's long-term goals, objectives, and strategic measures, which are the measurable human health and environmental results the Agency is working to achieve.[1] The *EPA Strategic Plan* acknowledges that the ability of communities to respond to changes in climate over the coming decades is critical to achieving many of the environmental outcomes the Agency is working towards. Consequently, the *EPA Strategic Plan* contains three strategic measures intended to promote the integration of climate adaptation planning into the Agency's activities.[2] The three strategic measures are focused on core Agency activities that influence its ability to fulfill its mission: (1) rule-making processes; (2) the distribution of financial resources and technical assistance; and (3) the development of science models and decision-support tools.

FY 2011-2015 Strategic Measures on Climate Adaptation

By 2015, EPA will account for climate change by integrating climate change science trend and scenario information into five rule-making processes to further EPA's mission, consistent with existing authorities (preference for one related to air quality, water quality, cleanup programs, and chemical safety.)

By 2015, EPA will build resilience to climate change by integrating considerations of climate change impacts and adaptive measures into five major grants, loan, contract, or technical assistance programs to further EPA's mission, consistent with existing authorities (preference for one related to air quality, water quality, cleanup programs, and scientific research).

By 2015, EPA will integrate climate change science trend and scenario information into five major scientific models and/or decision-support tools used in implementing Agency environmental management programs to further EPA's mission, consistent with existing authorities (preference for one related to air quality, water quality, cleanup programs, and chemical safety.)

The strategic measures are used by the Agency to design annual performance measures that are presented in EPA's *Annual Plans and Budgets*,

---

[1] U.S. Environmental Protection Agency, "FY 2011-2015 EPA Strategic Plan, Achieving Our Vision" (2011), 43, http://epa.gov/planandbudget/strategicplan.html.

[2] Performance measures can be used to provide accountability, as well as to guide decisions about program refinement and prioritization. They can be used to provide program managers and staff, and other external stakeholders, with valuable information about whether a project or program is meeting the desired goals. Measures can help identify when program goals are not being met and whether changes need to be made to meet those goals.

and to establish priorities in the annual National Program Manager (NPM) Guidance. The Agency then reports on its performance against these annual measures in the *Annual Performance Reports*.

In its *FY 2011-2015 Strategic Plan*, EPA emphasizes the importance of continuously evaluating activities based on their progress and performance, and acting on lessons learned. EPA is already using the three strategic measures pertaining to climate change adaptation to begin evaluating its actions on an ongoing basis. Through ongoing evaluation, the Agency will learn how to effectively mainstream climate adaptation planning into its activities. EPA will evaluate what worked and why, as well as what didn't work and why not. Based on the lessons, EPA will make adjustments to the way adaptation is integrated into its activities.

Importance of Program Evaluation

Because EPA programs and regions will be learning by experience as we integrate climate change adaptation into regulation, financial mechanisms and information tools, it will be essential to apply evaluation as a tool to better understand how well approaches work and how they can be improved upon.

4.2. New Performance Measures

Over time, the Agency will identify where its adaptation activities have or can have the greatest impact on protecting human health and the environment. However, it will be an ongoing challenge to measure the direct impact of EPA's adaptation planning activities on the resilience of its programs, and on the human health and environmental outcomes it is striving to attain. Metrics that enable one to attribute changes in resilience of environmental and human health outcomes to EPA's adaptation efforts, where this is possible, do not yet exist. Such metrics need to be developed over time.

Although the three existing strategic measures do not directly attribute changes in resilience of environmental and human health outcomes to EPA's adaptation efforts, they are focused on essential processes and outcomes (e.g., increased adaptive capacity gained through changes in knowledge and changes in behavior) that are important steps toward achieving the long-term goal of resilience to climate change.

As the Agency works to fulfill each of the three existing strategic measures, it might be possible to identify additional actions that must be taken to successfully attain the measures. For example, as EPA Program Offices integrate climate change adaptation into major rulemaking processes, they may discover that an effective approach is through the development of guidance for states and tribes authorized to implement Federal environmental programs. Identification of key steps like this might lead to the development of additional measures (e.g., numbers of states applying climate-related aspects of EPA guidance) for evaluating EPA's progress.

As EPA works with interested states and tribes to consider climate adaptation as they implement environmental programs, it could work with them to explore ways to measure changes in their adaptive capacity. Metrics could reflect *changes in knowledge* (e.g., number of partners taking formal training to increase their awareness of the importance of adaptation planning), *changes in behavior* (e.g., increases in the use of decision support tools to integrate climate adaptation planning into state and local planning activities such as infrastructure planning decisions), and *changes in state/condition* (e.g., changes in the ability of communities to withstand more frequent and intense storm events and avoid, for example, combined sewer overflow events).

Some of the pilot projects that will appear in Program and Regional Office Implementation Plans may also explore innovative ways for measuring changes in adaptive capacity through changes in knowledge, changes in behavior, or changes in state/condition. These will also inform the development of future Agency strategic measures.

EPA recognizes that the integration of climate adaptation planning into its programs, policies, rules, and operations will occur over time. This change will happen in stages and measures should reflect this evolution. The earliest changes in many programs will be changes in knowledge and awareness (e.g., increase in the awareness of EPA staff and their external partners of the relevance of adaptation planning to their programs). Building on this knowledge, they then will begin to change their behavior (e.g., increase their use of available decision support tools to integrate adaptation planning into their work). As programs mature, there will be evidence of more projects implemented as a result of increased attention to climate-related programmatic issues. Finally, in the long-term, adaptation planning efforts will lead to changes in condition (e.g., percentage of flood-prone communities that have increased their resilience to storm events) to directly support EPA's mission to protect human health and the environment.

Summary of Program Vulnerabilities to Climate Change Impacts by EPA Strategic Goal

Goal [a]	CLIMATE CHANGE IMPACTS [b]		EPA PROGRAMMATIC IMPACTS [c]		
	Climate Change Impact [d]	Likelihood of Impact [e]	Focus of Associated EPA Program	Likelihood EPA Program will be Affected by Impact [f]	Example of Risks if Program were Impacted
Goal 1: Taking Action on Climate Change and Improving Air Quality	• Increased tropospheric ozone pollution in certain regions	• Likely[1]	• Protecting public health and the environment by setting National Ambient Air Quality Standards (NAAQS) and implementing programs to help meet the standards	• High	• Could become more difficult to attain NAAQS for ozone in many areas with existing ozone problems
	• Increased frequency or intensity of wildfires	• Likely[2]	• Protecting public health and the environment by setting National Ambient Air Quality Standards (NAAQS) and implementing programs to help meet the standards	• Medium	• Could complicate Agency efforts to protect public health and the environment from risks posed by particulate matter (PM) pollution in areas affected by more frequent wildfires
	• Increasing extreme temperatures • Increasing heavy precipitation events	• Very Likely[3] • Likely[3]	• Protect public health by promoting healthy indoor environments through voluntary programs and guidance	• Medium	• Could increase public health risks, including risks for the young, the elderly, the chronically ill, and socioeconomically disadvantaged populations
	• Effects on the stratospheric ozone layer	• Likely[4]	• Restoring the stratospheric ozone layer • Preventing UV-related disease • Providing a smooth transition to safer alternatives	• High	• Unable to restore ozone concentrations to benchmark levels as quickly at some latitudes
	• Effects on response of ecosystems to atmospheric deposition of sulfur, nitrogen, and mercury	• Likely[5]	• Ecosystem protections from Agency emissions reduction programs	• Low	• Based on evolving research, could have consequences for the effectiveness of ecosystem protections under those programs

(Continued)

Goal [a]	CLIMATE CHANGE IMPACTS [b]		EPA PROGRAMMATIC IMPACTS [c]		
	Climate Change Impact [d]	Likelihood of Impact [e]	Focus of Associated EPA Program	Likelihood EPA Program will be Affected by Impact [f]	Example of Risks if Program were Impacted
Goal 2: Protecting America's Waters	• Increasing heavy precipitation events • Increasing intensity of hurricanes • Sea-level rise • Decreasing precipitation days and increasing drought intensity • Ocean acidification • Increased water temperatures	• Likely[3] • Likely[3] • Very likely[6] • Likely[7] • Certain[8] • Very Likely[9]	• Restoring and protecting watersheds, aquatic ecosystems and wetlands	• High	• Increased number of sewer overflows and wastewater bypasses, as well increased pollutant loads in runoff, fouling streams and threatening public health. • Challenges to coastal wetlands' ability to migrate. • Reduced streamflow, altering the aquatic environments and increasing impairments. • Continued stress on coral reefs. • Shifts in aquatic habitat will threaten the economic and cultural practices of tribal communities.
	• Increasing heavy precipitation events • Increasing intensity of hurricanes • Sea-level rise • Increasing intensity of hurricanes • Increasing flood risk	• Likely[3] • Likely[3] • Very likely[6] • Likely[3] • Likely[7]	• Drinking water, wastewater and stormwater infrastructure	• High	• Water infrastructure could be overwhelmed or damaged. • Drinking water intakes and wastewater outfalls could be affected. • Integrity of coastal water infrastructure systems could be put at increased risk. • Drinking water and wastewater utilities will need an 'all hazards' approach to planning for emergencies and extreme weather events. • Problems of safety as well as access to clean and safe water will be exacerbated for vulnerable and economically deprived communities.
	• Increased water temperatures • Increasing heavy precipitation events • Sea-level rise • Decreasing precipitation days and increasing drought intensity • Loss of snowpack	• Very likely[9] • Likely[3] • Very likely[6] • Likely[7] • Very likely[10]	• The quality and availability of safe drinking water	• Medium	• High water temperatures and increased stormwater runoff will increase the need for drinking water treatment, raising costs. • May cause saltwater intrusion in surface water and ground water, placing increased demands on drinking water treatment. • Water supplies may be affected, forcing communities to seek alternative sources. • Water demand may shift to underground aquifers or prompt development of reservoirs or underground storage of treated water, requiring EPA to ensure safety.

Goal [a]	CLIMATE CHANGE IMPACTS [b]		EPA PROGRAMMATIC IMPACTS [c]		
	Climate Change Impact [d]	Likelihood of Impact [e]	Focus of Associated EPA Program	Likelihood EPA Program will be Affected by Impact [f]	Example of Risks if Program were Impacted
Goal 3: Cleaning Up America's Communities & Advancing Sustainable Development	• Sea Level Rise • Increasing heavy precipitation events • Increasing risk of floods • Changes in temperature	• Very likely[6] • Likely[7] • Likely[7] • Very likely[3]	• Cleaning up Contaminated Sites and Waste Management	• Medium	• Increased risk of contaminant release from EPA Sites • May need to alter selected remedies to ensure protection. • Hazardous waste permitting may need to be updated to reflect climate change impacts.
	• Melting permafrost in Northern Regions	• Likely[10]	• Cleaning up Contaminated Sites and Waste Management	• High	• Increased risk of contaminant release at sites and potential impact to drinking water where permafrost was utilized as a containment remedy. • May need to implement new remedies to contain contaminants at sites previously protected by permafrost. •
	• Increasing intensity of hurricanes • Increasing heavy precipitation events • Increasing risk of floods	• Likely[3] • Likely[3] • Likely[7]	• Emergency Response	• Medium	• Increased need for emergency response. • Possible limitations to response capability due to staff and financial resource constraints. • Current waste management capacity, including interim capacity, may be insufficient to handle surges in necessary treatment and disposal of hazardous and municipal wastes, as well as mixed wastes generated from severe weather events.
Goal 4: Ensuring Safety of Chemicals & Preventing Pollution	• Increasing extreme temperatures • Increasing heavy precipitation events	• Very likely[3] • Likely[3]	• Protecting human health and ecosystems from chemical risks.	• Low	• Assure that chemical exposure models reflect changes in the environment • Changing in planting timing or location may affect the volume and timing of agricultural chemical use which could impact the appropriate risk management decisions.
Facilities and Operations	• Increased Water Temperatures • Decreasing precipitation days and increasing drought intensity	• Very likely[9] • Likely[7]	• Water usage at EPA facilities	• High	• Water temperatures impact research activities or cooling requirements. • Facilities could be located in areas with water shortages

(Continued)

Goal [a]	CLIMATE CHANGE IMPACTS [b]		EPA PROGRAMMATIC IMPACTS [c]		
	Climate Change Impact [d]	Likelihood of Impact [e]	Focus of Associated EPA Program	Likelihood EPA Program will be Affected by Impact [f]	Example of Risks if Program were Impacted
	• Increasing risk of floods • Increasing intensity of hurricanes • Sea level rise • Increasing extreme temperatures	• Likely[7] • Likely[3] • Very likely[6] • Very likely[3]	• Operations of Agency facilities, personnel safety, physical security, and emergency communications • Emergency management mission support (protective gear and acquisition)	• Medium	• Facilities in coastal or flood-prone areas • Personnel engaged in field work and vulnerable to extreme temperatures or events • Security, lighting and communication systems without backup power • Personnel and real property supporting emergency response and management

Footnotes for Summary of Climate Change Vulnerabilities to Climate Change Impacts by EPA Goal Table

[a] This table summarizes vulnerabilities by goal for four of the five goals in EPA's Strategic Plan. Goal 5 "Enforcing Environmental Laws" is not included in this table. Please note that the table also summarizes vulnerabilities to EPA facilities and operations; this is not part of the EPA Strategic Plan goal structure but is an important element of EPA's vulnerability assessment. Please see Section 2 of this document for a fuller discussion of impacts.

[b] Climate Change Impacts are based upon peer-reviewed scientific literature

[c] Programmatic Impacts are based upon EPA best professional judgment at this time.

[d] Impacts can vary by season and location.

[e] In general, the sources cited in this section use Intergovernmental Panel on Climate Change (IPCC) likelihood of outcome terminology where the term 'very likely' means 90-100% probability and the term 'likely' means 66-100% probability. For some impacts in the table, additional discussion on the likelihood term is provided in the associated footnote.

[f] High assumes the program will be affected by the impact; Medium assumes the program could be affected under some conditions by the impact; Low assumes that there is a potential for the program to be impacted or uncertainty currently exists as to the potential nature and extent of the impact. This assessment is based on best professional judgment within EPA at this time. Please note, this column does not reflect several important considerations. For example it does not distinguish timeframes (current, near-term, long-

term). It does not account for regional and local variations. And it does not reflect the priority of actions the agency may undertake now or in the future.

1) Denman, K.L., et al. (2007). Couplings Between Changes in the Climate System and Biogeochemistry. In: Climate Change 2007: The Physical Science Basis. Contribution of Working Group I to the Fourth Assessment Report of the Intergovernmental Panel on Climate Change [Solomon, S., D. Qin, M. Manning, Z. Chen, M. Marquis, K.B. Averyt, M. Tignor and H.L. Miller (eds.)]. Cambridge University Press, Cambridge, United Kingdom and New York, NY, USA.
2) C.B. Field et al., "North America," Chapter 14 in *Climate Change 2007: Impacts, Adaptation and Vulnerability*, Contribution of Working Group II to the Fourth Assessment Report of the Intergovernmental Panel on Climate Change, ed. M.L. Parry, O.F. Canziani, J.P. Palutikof, P.J. van der Linden and C.E. Hanson (Cambridge, United Kingdom and New York, NY, USA: Cambridge University Press, 2007).
3) IPCC, 2012: Summary for Policymakers. In: Managing the Risks of Extreme Events and Disasters to Advance Climate Change Adaptation [Field, C.B., V. Barros, T.F. Stocker, D. Qin, D.J. Dokken, K.L. Ebi, M.D. Mastrandrea,K.J. Mach, G.-K. Plattner, S.K. Allen, M. Tignor, and P.M. Midgley (eds.)]. A Special Report of Working Groups I and II of the Intergovernmental Panel on Climate Change. Cambridge University Press, Cambridge, UK, and New York, NY, USA, pp. 1-19.
4) World Meteorological Organization, *Scientific Assessment of Ozone Depletion: 2010*, Global Ozone Research and Monitoring Project—Report No. 52 (Geneva, Switzerland, 2011). Note: the word "expected" is used in the report to characterize projected climate change impacts on the stratospheric ozone layer. For purposes of this table the word "likely" has been used as a proxy for "expected."
5) Burns, D.A., Lynch, J.A., Cosby, B.J., Fenn, M.E., Baron, J.S., US EPA Clean Air Markets Div., 2011,National Acid Precipitation Assessment Program Report to Congress 2011: An Integrated Assessment, National Science and Technology Council, Washington, DC, p. 114.
6) IPCC, 2012: "It is very likely that mean sea level rise will contribute to upward trends in extreme coastal high water levels in the future."
7) USGCRP, 2009: *Global Climate Change Impacts in the United States.* Thomas R. Karl, Jerry M. Melillo, and Thomas C. Peterson (eds.). United States Global Change Research Program. Cambridge University Press, New York, NY, USA.

8) NRC, 2010: National Research Council of the National Academies, America's Climate Choices: Panel on Advancing the Science of Climate Change, 2010. p 41. "One of the most certain outcomes from increasing CO2 concentrations in the atmosphere is the acidification of the world's oceans." For purposes of this table, the term "certain" is used.

9) USGCRP, 2009: p. 46. [In the case of freshwater] "Increased air temperatures lead to higher water temperatures, which have already been detected in many streams, especially during low-flow periods." For the purposes of this table "very likely" is used.

10) Bates, B.C., Z.W. Kundzewicz, S. Wu and J.P. Palutikof, Eds., 2008: Climate Change and Water. Technical Paper of the Intergovernmental Panel on Climate Change, IPCC Secretariat, Geneva, p. 130.

End Notes for Part 1

[1] U.S. Environmental Protection Agency, *Policy Statement on Climate-Change Adaptation*, June 2, 2011, http://epa.gov/climatechange/Downloads/impacts-adaptation.

[2] Executive Order 13514 resulted in a process that requires every federal agency to submit annual progress reports to its sustainability and climate change adaptation plans.

[3] U.S. Environmental Protection Agency, *FY 2011-2015 EPA Strategic Plan, Achieving Our Vision* (2011), http://epa.gov/planandbudget/strategicplan.html.

[4] White House Council on Environmental Quality, *Progress Report on the Interagency Climate Change Adaptation Task Force: Recommended Actions in Support of a National Climate Change Adaptation Strategy* (Washington, DC, October 5, 2010).

End Notes for Part 2

[1] Thomas R. Karl, Jerry M. Melillo, and Thomas C. Peterson, eds., *Global Climate Change Impacts in the United States* (New York, NY: Cambridge University Press, 2009), http://downloads.globalchange.gov/usimpacts/pdfs/climate

[2] Thomas R. Karl, Jerry M. Melillo, and Thomas C. Peterson, eds., *Global Climate Change Impacts in the United States* (New York, NY: Cambridge University Press, 2009), http://downloads.globalchange.gov/usimpacts/pdfs/climate

[3] Thomas R. Karl, Jerry M. Melillo, and Thomas C. Peterson, eds., *Global Climate Change Impacts in the United States* (New York, NY: Cambridge University Press, 2009), http://downloads.globalchange.gov/usimpacts/pdfs/climate

[4] *Climate Change 2007: Impacts, Adaptation and Vulnerability*, Parry, M.L., Canziani, O. F., Palutikof, J. P., van der Linden, P. J., et. al., contribution of Working Group II to the Fourth Assessment Report of the Intergovernmental Panel on Climate Change 2007, "Summary for Policy Makers" (2007), 11-12.

[5] Thomas R. Karl, Jerry M. Melillo, and Thomas C. Peterson, eds., *Global Climate Change Impacts in the United States* (New York, NY: Cambridge University Press, 2009), 89–106, http://downloads.globalchange.gov/usimpacts/pdfs/climate

[6] U.S. Environmental Protection Agency, *Policy Statement on Climate-Change Adaptation*, June 2, 2011, http://epa.gov/climatechange/Downloads/impacts-adaptation

[7] While the discussion of vulnerabilities is organized by goal, some of the impacts of climate change are expected to cut across goals. For example, climate change may increase the number of weather-related emergencies to which the Agency is called to respond, potentially drawing staff from multiple locations and program areas. Staff participating in emergency response work would be unavailable to complete their normal program responsibilities. Climate change is also expected to affect many of the models that EPA programs and the states use to make environmental decisions.

[8] Denman, K.L., et al. (2007). Couplings Between Changes in the Climate System and Biogeochemistry. In: Climate Change 2007: The Physical Science Basis. Contribution of Working Group I to the Fourth Assessment Report of the Intergovernmental Panel on Climate Change [Solomon, S., D. Qin, M. Manning, Z. Chen, M. Marquis, K.B. Averyt, M. Tignor and H.L. Miller (eds.)]. Cambridge University Press, Cambridge, United Kingdom and New York, NY, USA.

[9] U.S. EPA (2009). Assessment of the Impacts of Global Change on Regional U.S. Air Quality: A Synthesis of Climate Change Impacts on Ground-Level Ozone. An Interim Report of the U.S. EPA Global Change Research Program. U.S. Environmental Protection Agency, Washington, DC, EPA/600/R-07/094F.

[10] Katsouyanni, K., et al.; HEI Health Review Committee (2009). Air pollution and health: a European and North American approach (APHENA). Research Report Health Effects Institute. 2009 Oct;(142):5-90.

[11] U.S. EPA. Air Quality Criteria for Ozone and Related Photochemical Oxidants (2006 Final). U.S. Environmental Protection Agency, Washington, DC, EPA/600/R-05/004aF-cF, 2006.

[12] Committee on Environment and Natural Resources, "Scientific Assessment of the Effects of Global Change on the United States" (Committee on Environment and Natural Resources of the National Science and Technology Council, U.S. Climate Change Science Program, 2008), http://www.climatescience.gov/Library/scientificassessment/Scientific-Assessment FINAL.pdf.

[13] C.B. Field et al., "North America," Chapter 14 in *Climate Change 2007: Impacts, Adaptation and Vulnerability*, Contribution of Working Group II to the Fourth Assessment Report of the Intergovernmental Panel on Climate Change, ed. M.L. Parry, O.F. Canziani, J.P. Palutikof, P.J. van der Linden and C.E. Hanson (Cambridge, United Kingdom and New York, NY, USA: Cambridge University Press, 2007).

[14] Institute of Medicine, *Climate Change, the Indoor Environment, and Health* (Washington, DC: The National Academies Press, 2011).

[15] World Meteorological Organization, *Scientific Assessment of Ozone Depletion: 2010*, Global Ozone Research and Monitoring Project—Report No. 52 (Geneva, Switzerland, 2011).

[16] Burns, D.A., Lynch, J.A., Cosby, B.J., Fenn, M.E., Baron, J.S., US EPA Clean Air Markets Div., 2011,National Acid Precipitation Assessment Program Report to Congress 2011: An Integrated Assessment, National Science and Technology Council, Washington, DC, 114 p.

[17] "Global Climate Change Impacts in the U.S."(2009), Water Sector, at: http://globalchange.gov/publications/reports/scientific-assessments/us-impacts/climate-

[18] There may be additional sites not included in this list that occur as a result of emergency response activities or unanticipated events.

[19] Conclusions from EPA consultations with the Federal Insecticide, Fungicide and Rodenticide Act (FIFRA) Scientific Advisory Panel (SAP), a panel of external experts established under FIFRA to provide the Agency with advice on pesticide-related science matters.

[20] U.S. Environmental Protection Agency, "FY 2011-2015 EPA Strategic Plan, Achieving our Vision" (2011), 43, http://epa.gov/planandbudget/strategicplan.html.

[21] Thomas R. Karl, Jerry M. Melillo, and Thomas C. Peterson, eds., *Global Climate Change Impacts in the United States* (New York, NY: Cambridge University Press, 2009), 89–106, http://downloads.globalchange.gov/usimpacts/pdfs/climate

[22] Thomas R. Karl, Jerry M. Melillo, and Thomas C. Peterson, eds., *Global Climate Change Impacts in the United States* (New York, NY: Cambridge University Press, 2009), 89–106, http://downloads.globalchange.gov/usimpacts/pdfs/climate

[23] G. Luber et al., "Climate Change and Extreme Heat Events," *Am J Prev Med* 35(5), August 2008.

[24] Thomas R. Karl, Jerry M. Melillo, and Thomas C. Peterson, eds., *Global Climate Change Impacts in the United States* (New York, NY: Cambridge University Press, 2009), 89–106, http://downloads.globalchange.gov/usimpacts/pdfs/climate

End Notes for Part 3

[1] U.S. Environmental Protection Agency, *Policy Statement on Climate-Change Adaptation*, June 2, 2011, http://epa.gov/climatechange/Downloads/impacts-adaptation

[2] U.S. Environmental Protection Agency, "FY 2011-2015 EPA Strategic Plan, Achieving Our Vision" (2011), 43, http://epa.gov/planandbudget/strategicplan.html.

[3] It will also be important, as EPA moves forward, to examine what impact rules being developed now could have in the future, taking climate into consideration.

[4] ICF Incorporated, "The Reg Map: Informal Rulemaking," 2003, http://www.reginfo.gov /public/reginfo/Regmap/regmap.pdf.

[5] National Research Council, *Informing Decisions in a Changing Climate*, Panel on Strategies and Methods for Climate-Related Decision Support, Committee on the Human Dimensions of Global Change, Division of Behavioral and Social Sciences and Education (Washington, DC: The National Academies Press, 2009).

[6] National Research Council, *Informing Decisions in a Changing Climate*, Panel on Strategies and Methods for Climate-Related Decision Support, Committee on the Human Dimensions of Global Change, Division of Behavioral and Social Sciences and Education (Washington, DC: The National Academies Press, 2009).

[7] S.L. Cutter and C. Finch, "Temporal and Spatial Changes in Social Vulnerability to Natural Hazards," *Proceedings of the National Academy of Science* 105(7) (2008), 2301-2306.

[8] In 2005 EPA started using the term life stages to refer to age-defined groups. The term life stage refers to a distinguishable time frame in an individual's life characterized by unique and relatively stable behavioral and/or physiological characteristics that are associated with development and growth. For example, EPA views childhood as a sequence of lifestages. There are other lifestages that may be important to consider when assessing human exposure and risk including, pregnancy, nursing, and middle and later years.

9 This includes the Office of Air and Radiation (OAR), Office of Water (OW), Office of Research and Development (ORD), Office of Solid Waste and Emergency Response (OSWER), Office of Enforcement and Compliance Assurance (OECA), Office of Chemical Safety and Pollution Prevention (OCSPP) and the Office of International and Tribal Affairs (OITA).

[10] National Water Program Strategy: Response to Climate Change. 2008 and 2012. Available at: http://www.epa.gov/water

In: Climate Change Adaptation Plan …
Editor: Cory Abrahams
ISBN: 978-1-63463-839-5
© 2015 Nova Science Publishers, Inc.

Chapter 2

EXECUTIVE ORDER 13653, PREPARING THE UNITED STATES FOR THE IMPACTS OF CLIMATE CHANGE*

Barack Obama

By the authority vested in me as President by the Constitution and the laws of the United States of America, and in order to prepare the Nation for the impacts of climate change by undertaking actions to enhance climate preparedness and resilience, it is hereby ordered as follows:

Section 1. Policy. The impacts of climate change -- including an increase in prolonged periods of excessively high temperatures, more heavy downpours, an increase in wildfires, more severe droughts, permafrost thawing, ocean acidification, and sea-level rise -- are already affecting communities, natural resources, ecosystems, economies, and public health across the Nation. These impacts are often most significant for communities that already face economic or health-related challenges, and for species and habitats that are already facing other pressures. Managing these risks requires deliberate preparation, close cooperation, and coordinated planning by the Federal Government, as well as by stakeholders, to facilitate Federal, State, local, tribal, private-sector, and nonprofit-sector efforts to improve climate preparedness and resilience; help safeguard our economy, infrastructure,

* This is an edited, reformatted and augmented version of an Executive Order issued by the White House, Office of the Press Secretary, November 1, 2013.

environment, and natural resources; and provide for the continuity of executive department and agency (agency) operations, services, and programs.

A foundation for coordinated action on climate change preparedness and resilience across the Federal Government was established by Executive Order 13514 of October 5, 2009 (Federal Leadership in Environmental, Energy, and Economic Performance), and the Interagency Climate Change Adaptation Task Force led by the Council on Environmental Quality (CEQ), the Office of Science and Technology Policy (OSTP), and the National Oceanic and Atmospheric Administration (NOAA). In addition, through the U.S. Global Change Research Program (USGCRP), established by section 103 of the Global Change Research Act of 1990 (15 U.S.C. 2933), and agency programs and activities, the Federal Government will continue to support scientific research, observational capabilities, and assessments necessary to improve our understanding of and response to climate change and its impacts on the Nation.

The Federal Government must build on recent progress and pursue new strategies to improve the Nation's preparedness and resilience. In doing so, agencies should promote: (1) engaged and strong partnerships and information sharing at all levels of government; (2) risk-informed decisionmaking and the tools to facilitate it; (3) adaptive learning, in which experiences serve as opportunities to inform and adjust future actions; and (4) preparedness planning.

Sec. 2. Modernizing Federal Programs to Support Climate Resilient Investment. (a) To support the efforts of regions, States, local communities, and tribes, all agencies, consistent with their missions and in coordination with the Council on Climate Preparedness and Resilience (Council) established in section 6 of this order, shall:

i. identify and seek to remove or reform barriers that discourage investments or other actions to increase the Nation's resilience to climate change while ensuring continued protection of public health and the environment;
ii. reform policies and Federal funding programs that may, perhaps unintentionally, increase the vulnerability of natural or built systems, economic sectors, natural resources, or communities to climate change related risks;
iii. identify opportunities to support and encourage smarter, more climate-resilient investments by States, local communities, and tribes, including by providing incentives through agency guidance, grants,

technical assistance, performance measures, safety considerations, and other programs, including in the context of infrastructure development as reflected in Executive Order 12893 of January 26, 1994 (Principles for Federal Infrastructure Investments), my memorandum of August 31, 2011 (Speeding Infrastructure Development through More Efficient and Effective Permitting and Environmental Review), Executive Order 13604 of March 22, 2012 (Improving Performance of Federal Permitting and Review of Infrastructure Projects), and my memorandum of May 17, 2013 (Modernizing Federal Infrastructure Review and Permitting Regulations, Policies, and Procedures); and

iv. report on their progress in achieving the requirements identified above, including accomplished and planned milestones, in the Agency Adaptation Plans developed pursuant to section 5 of this order.

(b) In carrying out this section, agencies should also consider the recommendations of the State, Local, and Tribal Leaders Task Force on Climate Preparedness and Resilience (Task Force) established in section 7 of this order and the National Infrastructure Advisory Council established by Executive Order 13231 of October 16, 2001 (Critical Infrastructure Protection in the Information Age), and continued through Executive Order 13652 of September 30, 2013 (Continuance of Certain Federal Advisory Committees).

(c) Interagency groups charged with coordinating and modernizing Federal processes related to the development and integration of both man-made and natural infrastructure, evaluating public health and social equity issues, safeguarding natural resources, and other issues impacted by climate change -- including the Steering Committee on Federal Infrastructure Permitting and Review Process Improvement established by Executive Order 13604, the Task Force on Ports established on July 19, 2012, the Interagency Working Group on Coordination of Domestic Energy Development and Permitting in Alaska established by Executive Order 13580 of July 12, 2011, and the Federal Interagency Working Group on Environmental Justice established by Executive Order 12898 of February 11, 1994 -- shall be responsible for ensuring that climate change related risks are accounted for in such processes and shall work with agencies in meeting the requirements set forth in subsections (a) and (b) of this section.

Sec. 3. Managing Lands and Waters for Climate Preparedness and Resilience. Within 9 months of the date of this order and in coordination with the efforts described in section 2 of this order, the heads of the Departments of Defense, the Interior, and Agriculture, the Environmental Protection Agency,

NOAA, the Federal Emergency Management Agency, the Army Corps of Engineers, and other agencies as recommended by the Council established in section 6 of this order shall work with the Chair of CEQ and the Director of the Office of Management and Budget (OMB) to complete an inventory and assessment of proposed and completed changes to their land- and water-related policies, programs, and regulations necessary to make the Nation's watersheds, natural resources, and ecosystems, and the communities and economies that depend on them, more resilient in the face of a changing climate. Further, recognizing the many benefits the Nation's natural infrastructure provides, agencies shall, where possible, focus on program and policy adjustments that promote the dual goals of greater climate resilience and carbon sequestration, or other reductions to the sources of climate change. The assessment shall include a timeline and plan for making changes to policies, programs, and regulations. Agencies shall build on efforts already completed or underway as outlined in agencies' Adaptation Plans, as discussed in section 5 of this order, as well as recent interagency climate adaptation strategies such as the *National Action Plan: Priorities for Managing Freshwater Resources in a Changing Climate*, released October 28, 2011; the *National Fish, Wildlife and Plants Climate Adaptation Strategy*, released March 26, 2013; and the *National Ocean Policy Implementation Plan*, released April 16, 2013.

Sec. 4. Providing Information, Data, and Tools for Climate Change Preparedness and Resilience. (a) In support of Federal, regional, State, local, tribal, private-sector and nonprofit-sector efforts to prepare for the impacts of climate change, the Departments of Defense, the Interior, Agriculture, Commerce, Health and Human Services, Housing and Urban Development, Transportation, Energy, and Homeland Security, the Environmental Protection Agency, the National Aeronautics and Space Administration, and any other agencies as recommended by the Council established in section 6 of this order, shall, supported by USGCRP, work together to develop and provide authoritative, easily accessible, usable, and timely data, information, and decision-support tools on climate preparedness and resilience.

(b) As part of the broader open data policy, CEQ and OSTP, in collaboration with OMB and consistent with Executive Order 13642 of May 9, 2013 (Making Open and Machine Readable the New Default for Government Information), shall oversee the establishment of a web-based portal on "Data.gov" and work with agencies on identifying, developing, and integrating data and tools relevant to climate issues and decisionmaking. Agencies shall coordinate their work on these data and tools with relevant interagency councils and committees such as the National Science and Technology

Council and those that support the implementation of Presidential Policy Directive-21 of February 12, 2013 (Critical Infrastructure Security and Resilience).

Sec. 5. Federal Agency Planning for Climate Change Related Risk. (a) Consistent with Executive Order 13514, agencies have developed Agency Adaptation Plans and provided them to CEQ and OMB. These plans evaluate the most significant climate change related risks to, and vulnerabilities in, agency operations and missions in both the short and long term, and outline actions that agencies will take to manage these risks and vulnerabilities. Building on these efforts, each agency shall develop or continue to develop, implement, and update comprehensive plans that integrate consideration of climate change into agency operations and overall mission objectives and submit those plans to CEQ and OMB for review. Each Agency Adaptation Plan shall include:

(i) identification and assessment of climate change related impacts on and risks to the agency's ability to accomplish its missions, operations, and programs;
(ii) a description of programs, policies, and plans the agency has already put in place, as well as additional actions the agency will take, to manage climate risks in the near term and build resilience in the short and long term;
(iii) a description of how any climate change related risk identified pursuant to paragraph (i) of this subsection that is deemed so significant that it impairs an agency's statutory mission or operation will be addressed, including through the agency's existing reporting requirements;
(iv) a description of how the agency will consider the need to improve climate adaptation and resilience, including the costs and benefits of such improvement, with respect to agency suppliers, supply chain, real property investments, and capital equipment purchases such as updating agency policies for leasing, building upgrades, relocation of existing facilities and equipment, and construction of new facilities; and
(v) a description of how the agency will contribute to coordinated interagency efforts to support climate preparedness and resilience at all levels of government, including collaborative work across agencies' regional offices and hubs, and through coordinated

development of information, data, and tools, consistent with section 4 of this order.

(b) Agencies will report on progress made on their Adaptation Plans, as well as any updates made to the plans, through the annual Strategic Sustainability Performance Plan process. Agencies shall regularly update their Adaptation Plans, completing the first update within 120 days of the date of this order, with additional regular updates thereafter due not later than 1 year after the publication of each quadrennial National Climate Assessment report required by section 106 of the Global Change Research Act of 1990 (15 U.S.C. 2936).

Sec. 6. Council on Climate Preparedness and Resilience.

(a) Establishment. There is established an interagency Council on Climate Preparedness and Resilience (Council).

(b) Membership. The Council shall be co-chaired by the Chair of CEQ, the Director of OSTP, and the Assistant to the President for Homeland Security and Counterterrorism. In addition, the Council shall include senior officials (Deputy Secretary or equivalent officer) from:

(i) the Department of State;
(ii) the Department of the Treasury;
(iii) the Department of Defense;
(iv) the Department of Justice;
(v) the Department of the Interior;
(vi) the Department of Agriculture;
(vii) the Department of Commerce;
(viii) the Department of Labor;
(ix) the Department of Health and Human Services;
(x) the Department of Housing and Urban Development;
(xi) the Department of Transportation;
(xii) the Department of Energy;
(xiii) the Department of Education;
(xiv) the Department of Veterans Affairs;
(xv) the Department of Homeland Security;
(xvi) the United States Agency for International Development;
(xvii) the Army Corps of Engineers;
(xviii) the Environmental Protection Agency;
(xix) the General Services Administration;
(xx) the Millennium Challenge Corporation;

(xxi) the National Aeronautics and Space Administration;
(xxii) the U.S. Small Business Administration;
(xxiii) the Corporation for National and Community Service;
(xxiv) the Office of the Director of National Intelligence;
(xxv) the Council of Economic Advisers;
(xxvi) the National Economic Council;
(xxvii) the Domestic Policy Council;
(xxviii) the Office of Management and Budget;
(xxix) the White House Office of Public Engagement and Intergovernmental Affairs;
(xxx) the United States Trade Representative; and
(xxxi) such agencies or offices as the President or Co-Chairs shall designate.

(c) Administration. CEQ shall provide administrative support and additional resources, as appropriate, for the Council to the extent permitted by law and within existing appropriations. Agencies shall assist and provide information to the Council, consistent with applicable law, as may be necessary to carry out its functions. Each agency shall bear its own expenses for participating in the Council.

(d) Council Structure. The Co-Chairs shall designate a subset of members of the Council to serve on a Steering Committee, which shall help determine priorities and strategic direction for the Council. The Co-Chairs and Steering Committee may establish working groups as needed, and may recharter working groups of the Interagency Climate Change Adaptation Task Force, as appropriate.

(e) Mission and Function of the Council. The Council shall work across agencies and offices, and in partnership with State, local, and tribal governments (as well as the Task Force established in section 7 of this order), academic and research institutions, and the private and nonprofit sectors to:

(i) develop, recommend, coordinate interagency efforts on, and track implementation of priority Federal Government actions related to climate preparedness and resilience;

(ii) support regional, State, local, and tribal action to assess climate change related vulnerabilities and cost-effectively increase climate preparedness and resilience of communities, critical economic sectors, natural and built infrastructure, and natural resources, including through the activities as outlined in sections 2 and 3 of this order;

(iii) facilitate the integration of climate science in policies and planning of government agencies and the private sector, including by promoting the development of innovative, actionable, and accessible Federal climate change related information, data, and tools at appropriate scales for decisionmakers and deployment of this information through a Government-wide web-based portal, as described in section 4 of this order; and

(iv) such other functions as may be decided by the Co-Chairs, including implementing, as appropriate, the recommendations of the Task Force established in section 7 of this order.

(f) Termination of the Interagency Climate Change Adaptation Task Force. The Interagency Climate Change Adaptation Task Force (Adaptation Task Force), established in 2009, created the framework for coordinated Federal action on climate preparedness and resilience, driving agency-level planning and action. The Adaptation Task Force shall terminate no later than 30 days after the first meeting of the Council, which shall continue and build upon the Adaptation Task Force's work.

Sec. 7. State, Local, and Tribal Leaders Task Force on Climate Preparedness and Resilience.

(a) Establishment. To inform Federal efforts to support climate preparedness and resilience, there is established a State, Local, and Tribal Leaders Task Force on Climate Preparedness and Resilience (Task Force).

(b) Membership. The Task Force shall be co-chaired by the Chair of CEQ and the Director of the White House Office of Intergovernmental Affairs. In addition, its members shall be such elected State, local, and tribal officials as may be invited by the Co-Chairs to participate. Members of the Task Force, acting in their official capacity, may designate employees with authority to act on their behalf.

(c) Mission and Function. Within 1 year of the date of this order, the Task Force shall provide, through its Co-Chairs, recommendations to the President and the Council for how the Federal Government can:

(i) remove barriers, create incentives, and otherwise modernize Federal programs to encourage investments, practices, and partnerships that facilitate increased resilience to climate impacts, including those associated with extreme weather;

(ii) provide useful climate preparedness tools and actionable information for States, local communities, and tribes, including through interagency collaboration as described in section 6 of this order; and

(iii) otherwise support State, local, and tribal preparedness for and resilience to climate change.

(d) Sunset. The Task Force shall terminate no later than 6 months after providing its recommendations.

Sec. 8. Definitions. As used in this order:

(a) "preparedness" means actions taken to plan, organize, equip, train, and exercise to build, apply, and sustain the capabilities necessary to prevent, protect against, ameliorate the effects of, respond to, and recover from climate change related damages to life, health, property, livelihoods, ecosystems, and national security;

(b) "adaptation" means adjustment in natural or human systems in anticipation of or response to a changing environment in a way that effectively uses beneficial opportunities or reduces negative effects; and

(c) "resilience" means the ability to anticipate, prepare for, and adapt to changing conditions and withstand, respond to, and recover rapidly from disruptions.

Sec. 9. General Provisions. (a) Nothing in this order shall be construed to impair or otherwise affect:

(i) the authority granted by law to an executive department, agency, or the head thereof; or

(ii) the functions of the Director of OMB relating to budgetary, administrative, or legislative proposals.

(b) This order shall be implemented consistent with U.S. obligations under international agreements and applicable U.S. law, and be subject to the availability of appropriations.

(c) This order is not intended to, and does not, create any right or benefit, substantive or procedural, enforceable at law or in equity by any party against the United States, its departments, agencies, or entities, its officers, employees, or agents, or any other person.

BARACK OBAMA

In: Climate Change Adaptation Plan ...
Editor: Cory Abrahams

ISBN: 978-1-63463-839-5
© 2015 Nova Science Publishers, Inc.

Chapter 3

THE PRESIDENT'S CLIMATE ACTION PLAN*

Executive Office of the President

PRESIDENT OBAMA'S CLIMATE ACTION PLAN

"We, the people, still believe that our obligations as Americans are not just to ourselves, but to all posterity. We will respond to the threat of climate change, knowing that the failure to do so would betray our children and future generations. Some may still deny the overwhelming judgment of science, but none can avoid the devastating impact of raging fires and crippling drought and more powerful storms.

The path towards sustainable energy sources will be long and sometimes difficult. But America cannot resist this transition, we must lead it. We cannot cede to other nations the technology that will power new jobs and new industries, we must claim its promise. That's how we will maintain our economic vitality and our national treasure -- our forests and waterways, our croplands and snow-capped peaks. That is how we will preserve our planet, commanded to our care by God. That's what will lend meaning to the creed our fathers once declared."

-- President Obama, Second Inaugural Address, January 2013

* This is an edited, reformatted and augmented version of an Executive Office of the President publication, dated June 2013.

The Case for Action

While no single step can reverse the effects of climate change, we have a moral obligation to future generations to leave them a planet that is not polluted and damaged. Through steady, responsible action to cut carbon pollution, we can protect our children's health and begin to slow the effects of climate change so that we leave behind a cleaner, more stable environment.

In 2009, President Obama made a pledge that by 2020, America would reduce its greenhouse gas emissions in the range of 17 percent below 2005 levels if all other major economies agreed to limit their emissions as well. Today, the President remains firmly committed to that goal and to building on the progress of his first term to help put us and the world on a sustainable long-term trajectory. Thanks in part to the Administration's success in doubling America's use of wind, solar, and geothermal energy and in establishing the toughest fuel economy standards in our history, we are creating new jobs, building new industries, and reducing dangerous carbon pollution which contributes to climate change. In fact, last year, carbon emissions from the energy sector fell to the lowest level in two decades. At the same time, while there is more work to do, we are more energy secure than at any time in recent history. In 2012, America's net oil imports fell to the lowest level in 20 years and we have become the world's leading producer of natural gas – the cleanest-burning fossil fuel.

While this progress is encouraging, climate change is no longer a distant threat – we are already feeling its impacts across the country and the world. Last year was the warmest year ever in the contiguous United States and about one-third of all Americans experienced 10 days or more of 100-degree heat. The 12 hottest years on record have all come in the last 15 years. Asthma rates have doubled in the past 30 years and our children will suffer more asthma attacks as air pollution gets worse. And increasing floods, heat waves, and droughts have put farmers out of business, which is already raising food prices dramatically.

These changes come with far-reaching consequences and real economic costs. Last year alone, there were 11 different weather and climate disaster events with estimated losses exceeding $1 billion each across the United States. Taken together, these 11 events resulted in over $110 billion in estimated damages, which would make it the second-costliest year on record.

In short, America stands at a critical juncture. Today, President Obama is putting forward a broad-based plan to cut the carbon pollution that causes climate change and affects public health. Cutting carbon pollution will help spark business innovation to modernize our power plants, resulting in cleaner forms of American-made energy that will create good jobs and cut our dependence on foreign oil. Combined with the Administration's other actions to increase the efficiency of our cars and household appliances, the President's plan will reduce the amount of energy consumed by American families, cutting down on their gas and utility bills. The plan, which consists of a wide variety of executive actions, has three key pillars:

1. *Cut Carbon Pollution in America:* In 2012, U.S. carbon emissions fell to the lowest level in two decades even as the economy continued to grow. To build on this progress, the Obama Administration is putting in place tough new rules to cut carbon pollution – just like we have for other toxins like mercury and arsenic – so we protect the health of our children and move our economy toward American-made clean energy sources that will create good jobs and lower home energy bills.
2. *Prepare the United States for the Impacts of Climate Change:* Even as we take new steps to reduce carbon pollution, we must also prepare for the impacts of a changing climate that are already being felt across the country. Moving forward, the Obama Administration will help state and local governments strengthen our roads, bridges, and shorelines so we can better protect people's homes, businesses and way of life from severe weather.
3. *Lead International Efforts to Combat Global Climate Change and Prepare for its Impacts:* Just as no country is immune from the impacts of climate change, no country can meet this challenge alone. That is why it is imperative for the United States to couple action at home with leadership internationally. America must help forge a truly global solution to this global challenge by galvanizing international action to significantly reduce emissions (particularly among the major emitting countries), prepare for climate impacts, and drive progress through the international negotiations.

Climate change represents one of our greatest challenges of our time, but it is a challenge uniquely suited to America's strengths. Our scientists will design new fuels, and our farmers will grow them. Our engineers to devise new sources of energy, our workers will build them, and our businesses will sell them. All of us will need to do our part. If we embrace this challenge, we will not just create new jobs and new industries and keep America on the cutting edge; we will save lives, protect and preserve our treasured natural resources, cities, and coastlines for future generations.

What follows is a blueprint for steady, responsible national and international action to slow the effects of climate change so we leave a cleaner, more stable environment for future generations. It highlights progress already set in motion by the Obama Administration to advance these goals and sets forth new steps to achieve them.

CUT CARBON POLLUTION IN AMERICA

In 2009, President Obama made a commitment to reduce U.S. greenhouse gas emissions in the range of 17 percent below 2005 levels by 2020. The President remains firmly committed to achieving that goal. While there is more work to do, the Obama Administration has already made significant progress by doubling generation of electricity from wind, solar, and geothermal, and by establishing historic new fuel economy standards. Building on these achievements, this document outlines additional steps the Administration will take – in partnership with states, local communities, and the private sector – to continue on a path to meeting the President's 2020 goal.

I. Deploying Clean Energy

Cutting Carbon Pollution from Power Plants: Power plants are the largest concentrated source of emissions in the United States, together accounting for roughly one-third of all domestic greenhouse gas emissions. We have already set limits for arsenic, mercury, and lead, but there is no federal rule to prevent power plants from releasing as much carbon pollution as they want. Many states, local governments, and companies have taken steps to move to cleaner electricity sources. More than 35 states have renewable energy targets in place, and more than 25 have set energy efficiency targets.

Despite this progress at the state level, there are no federal standards in place to reduce carbon pollution from power plants. In April 2012, as part of a continued effort to modernize our electric power sector, the Obama Administration proposed a carbon pollution standard for new power plants. The Environmental Protection Agency's proposal reflects and reinforces the ongoing trend towards cleaner technologies, with natural gas increasing its share of electricity generation in recent years, principally through market forces and renewables deployment growing rapidly to account for roughly half of new generation capacity installed in 2012.

With abundant clean energy solutions available, and building on the leadership of states and local governments, we can make continued progress in reducing power plant pollution to improve public health and the environment while supplying the reliable, affordable power needed for economic growth. By doing so, we will continue to drive American leadership in clean energy technologies, such as efficient natural gas, nuclear, renewables, and clean coal technology.

To accomplish these goals, President Obama is issuing a Presidential Memorandum directing the Environmental Protection Agency to work expeditiously to complete carbon pollution standards for both new and existing power plants. This work will build on the successful first-term effort to develop greenhouse gas and fuel economy standards for cars and trucks. In developing the standards, the President has asked the Environmental Protection Agency to build on state leadership, provide flexibility, and take advantage of a wide range of energy sources and technologies including many actions in this plan.

Promoting American Leadership in Renewable Energy: During the President's first term, the United States more than doubled generation of electricity from wind, solar, and geothermal sources. To ensure America's continued leadership position in clean energy, President Obama has set a goal to double renewable electricity generation once again by 2020. In order to meet this ambitious target, the Administration is announcing a number of new efforts in the following key areas:

- *Accelerating Clean Energy Permitting:* In 2012 the President set a goal to issue permits for 10 gigawatts of renewables on public lands by the end of the year. The Department of the Interior achieved this goal ahead of schedule and the President has directed it to permit an additional 10 gigawatts by 2020. Since 2009, the Department of Interior has approved 25 utility-scale solar facilities, nine wind farms, and 11 geothermal plants, which will provide enough electricity to power 4.4 million homes and support an estimated 17,000 jobs. The Administration is also taking steps to encourage the development of hydroelectric power at existing dams. To develop and demonstrate improved permitting procedures for such projects, the Administration will designate the Red Rock Hydroelectric Plant on the Des Moines River in Iowa to participate in its Infrastructure Permitting Dashboard for high-priority projects. Also, the Department of Defense – the single largest consumer of energy in the United States – is committed to deploying 3 gigawatts of renewable energy on military installations, including solar, wind, biomass, and geothermal, by 2025. In addition, federal agencies are setting a new goal of reaching 100 megawatts of installed renewable capacity across the federally subsidized housing stock by 2020. This effort will include conducting a survey of current projects in order to track progress and facilitate the sharing of best practices.

- *Expanding and Modernizing the Electric Grid:* Upgrading the country's electric grid is critical to our efforts to make electricity more reliable, save consumers money on their energy bills, and promote clean energy sources. To advance these important goals, President Obama signed a Presidential Memorandum this month that directs federal agencies to streamline the siting, permitting and review process for transmission projects across federal, state, and tribal governments.

Unlocking Long-Term Investment in Clean Energy Innovation: The Fiscal Year 2014 Budget continues the President's commitment to keeping the United States at the forefront of clean energy research, development, and deployment by increasing funding for clean energy technology across all agencies by 30 percent, to approximately $7.9 billion. This includes investment in a range of energy technologies, from advanced biofuels and emerging nuclear technologies – including small modular reactors – to clean coal. To continue America's leadership in clean energy innovation, the Administration will also take the following steps:

- *Spurring Investment in Advanced Fossil Energy Projects*: In the coming weeks, the Department of Energy will issue a Federal Register Notice announcing a draft of a solicitation that would make up to $8 billion in (self-pay) loan guarantee authority available for a wide array of advanced fossil energy projects under its Section 1703 loan guarantee program. This solicitation is designed to support investments in innovative technologies that can cost-effectively meet financial and policy goals, including the avoidance, reduction, or sequestration of anthropogenic emissions of greenhouse gases. The proposed solicitation will cover a broad range of advanced fossil energy projects. Reflecting the Department's commitment to continuous improvement in program management, it will take comment on the draft solicitation, with a plan to issue a final solicitation by the fall of 2013.

- *Instituting a Federal Quadrennial Energy Review*: Innovation and new sources of domestic energy supply are transforming the nation's energy marketplace, creating economic opportunities at the same time they raise environmental challenges. To ensure that federal energy policy meets our economic, environmental, and security goals in this changing landscape, the Administration will conduct a Quadrennial Energy Review which will be led by the White House Domestic Policy Council and Office of Science and Technology Policy, supported by a Secretariat established at the Department of Energy, and involving the robust engagement of federal agencies and outside stakeholders. This first-ever review will focus on infrastructure challenges, and will identify the threats, risks, and opportunities for U.S. energy and climate security, enabling the federal government to translate policy goals into a set of analytically based, clearly articulated, sequenced and integrated actions, and proposed investments over a four-year planning horizon.

II. Building a 21st-Century Transportation Sector

Increasing Fuel Economy Standards: Heavy-duty vehicles are currently the second largest source of greenhouse gas emissions within the transportation sector. In 2011, the Obama Administration finalized the first-ever fuel economy standards for Model Year 2014-2018 for heavy-duty trucks, buses, and vans. These standards will reduce greenhouse gas emissions by approximately 270 million metric tons and save 530 million barrels of oil. During the President's second term, the Administration will once again partner with industry leaders and other key stakeholders to develop post-2018 fuel economy standards for heavy-duty vehicles to further reduce fuel consumption through the application of advanced cost-effective technologies and continue efforts to improve the efficiency of moving goods across the United States.

The Obama Administration has already established the toughest fuel economy standards for passenger vehicles in U.S. history. These standards require an average performance equivalent of 54.5 miles per gallon by 2025, which will save the average driver more than $8,000 in fuel costs over the lifetime of the vehicle and eliminate six billion metric tons of carbon pollution – more than the United States emits in an entire year.

Developing and Deploying Advanced Transportation Technologies: Biofuels have an important role to play in increasing our energy security, fostering rural economic development, and reducing greenhouse gas emissions from the transportation sector. That is why the Administration supports the Renewable Fuels Standard, and is investing in research and development to help bring next-generation biofuels on line. For example, the United States Navy and Departments of Energy and Agriculture are working with the private sector to accelerate the development of cost-competitive advanced biofuels for use by the military and commercial sectors. More broadly, the Administration will continue to leverage partnerships between the private and public sectors to deploy cleaner fuels, including advanced batteries and fuel cell technologies, in every transportation mode. The Department of Energy's eGallon informs drivers about electric car operating costs in their state – the national average is only $1.14 per gallon of gasoline equivalent, showing the promise for consumer pocketbooks of electric-powered vehicles. In addition, in the coming months, the Department of Transportation will work with other agencies to further explore strategies for integrating alternative fuel vessels into the U.S. flag fleet. Further, the Administration will continue to work with states, cities and towns through the Department of Transportation, the Department of Housing and Urban Development, and the Environmental Protection Agency to improve transportation options, and lower transportation costs while protecting the environment in communities nationwide.

III. Cutting Energy Waste in Homes, Businesses, and Factories

Reducing Energy Bills for American Families and Businesses: Energy efficiency is one of the clearest and most cost-effective opportunities to save families money, make our businesses more competitive, and reduce greenhouse gas emissions. In the President's first term, the Department of Energy and the Department of Housing and Urban Development completed efficiency upgrades in more than one million homes, saving many families more than $400 on their heating and cooling bills in the first year alone. The Administration will take a range of new steps geared towards achieving President Obama's goal of doubling energy productivity by 2030 relative to 2010 levels:

- *Establishing a New Goal for Energy Efficiency Standards*: In President Obama's first term, the Department of Energy established new minimum efficiency standards for dishwashers, refrigerators, and many other products. Through 2030, these standards will cut consumers' electricity bills by hundreds of billions of dollars and save enough electricity to power more than 85 million homes for two years. To build on this success, the Administration is setting a new goal: Efficiency standards for appliances and federal buildings set in the first and second terms combined will reduce carbon pollution by at least 3 billion metric tons cumulatively by 2030 – equivalent to nearly one-half of the carbon pollution from the entire U.S. energy sector for one year – while continuing to cut families' energy bills.
- *Reducing Barriers to Investment in Energy Efficiency*: Energy efficiency upgrades bring significant cost savings, but upfront costs act as a barrier to more widespread investment. In response, the Administration is committing to a number of new executive actions. As soon as this fall, the Department of Agriculture's Rural Utilities Service will finalize a proposed update to its Energy Efficiency and Conservation Loan Program to provide up to $250 million for rural utilities to finance efficiency investments by businesses and homeowners across rural America. The Department is also streamlining its Rural Energy for America program to provide grants and loan guarantees directly to agricultural producers and rural small businesses for energy efficiency and renewable energy systems.

  In addition, the Department of Housing and Urban Development's efforts include a $23 million Multifamily Energy Innovation Fund designed to enable affordable housing providers, technology firms, academic institutions, and philanthropic organizations to test new approaches to deliver cost-effective residential energy. In order to advance ongoing efforts and bring stakeholders together, the Federal Housing Administration will convene representatives of the lending community and other key stakeholders for a mortgage roundtable in July to identify options for factoring energy efficiency into the mortgage underwriting and appraisal process upon sale or refinancing of new or existing homes.
- *Expanding the President's Better Buildings Challenge*: The Better Buildings Challenge, focused on helping American commercial and industrial buildings become at least 20 percent more energy efficient by 2020, is already showing results. More than 120 diverse organizations, representing over 2 billion square feet are on track to

meet the 2020 goal: cutting energy use by an average 2.5 percent annually, equivalent to about $58 million in energy savings per year. To continue this success, the Administration will expand the program to multifamily housing – partnering both with private and affordable building owners and public housing agencies to cut energy waste. In addition, the Administration is launching the Better Buildings Accelerators, a new track that will support and encourage adoption of State and local policies to cut energy waste, building on the momentum of ongoing efforts at that level.

IV. Reducing Other Greenhouse Gas Emissions

Curbing Emissions of Hydrofluorocarbons: Hydrofluorocarbons (HFCs), which are primarily used for refrigeration and air conditioning, are potent greenhouse gases. In the United States, emissions of HFCs are expected to nearly triple by 2030, and double from current levels of 1.5 percent of greenhouse gas emissions to 3 percent by 2020.

To reduce emissions of HFCs, the United States can and will lead both through international diplomacy as well as domestic actions. In fact, the Administration has already acted by including a flexible and powerful incentive in the fuel economy and carbon pollution standards for cars and trucks to encourage automakers to reduce HFC leakage and transition away from the most potent HFCs in vehicle air conditioning systems. Moving forward, the Environmental Protection Agency will use its authority through the Significant New Alternatives Policy Program to encourage private sector investment in low-emissions technology by identifying and approving climate-friendly chemicals while prohibiting certain uses of the most harmful chemical alternatives. In addition, the President has directed his Administration to purchase cleaner alternatives to HFCs whenever feasible and transition over time to equipment that uses safer and more sustainable alternatives.

Reducing Methane Emissions: Curbing emissions of methane is critical to our overall effort to address global climate change. Methane currently accounts for roughly 9 percent of domestic greenhouse gas emissions and has a global warming potential that is more than 20 times greater than carbon dioxide. Notably, since 1990, methane emissions in the United States have decreased by 8 percent. This has occurred in part through partnerships with industry, both at home and abroad, in which we have demonstrated that we have the technology to deliver emissions reductions that benefit both our economy and the environment. To achieve additional progress, the Administration will:

- *Developing an Interagency Methane Strategy:* The Environmental Protection Agency and the Departments of Agriculture, Energy, Interior, Labor, and Transportation will develop a comprehensive, interagency methane strategy. The group will focus on assessing current emissions data, addressing data gaps, identifying technologies and best practices for reducing emissions, and identifying existing authorities and incentive-based opportunities to reduce methane emissions.
- *Pursuing a Collaborative Approach to Reducing Emissions:* Across the economy, there are multiple sectors in which methane emissions can be reduced, from coal mines and landfills to agriculture and oil and gas development. For example, in the agricultural sector, over the last three years, the Environmental Protection Agency and the Department of Agriculture have worked with the dairy industry to increase the adoption of methane digesters through loans, incentives, and other assistance. In addition, when it comes to the oil and gas sector, investments to build and upgrade gas pipelines will not only put more Americans to work, but also reduce emissions and enhance economic productivity. For example, as part of the Administration's effort to improve federal permitting for infrastructure projects, the interagency Bakken Federal Executive Group is working with industry, as well as state and tribal agencies, to advance the production of oil and gas in the Bakken while helping to reduce venting and flaring. Moving forward, as part of the effort to develop an interagency methane strategy, the Obama Administration will work collaboratively with state governments, as well as the private sector, to reduce emissions across multiple sectors, improve air quality, and achieve public health and economic benefits.

Preserving the Role of Forests in Mitigating Climate Change: America's forests play a critical role in addressing carbon pollution, removing nearly 12 percent of total U.S. greenhouse gas emissions each year. In the face of a changing climate and increased risk of wildfire, drought, and pests, the capacity of our forests to absorb carbon is diminishing. Pressures to develop forest lands for urban or agricultural uses also contribute to the decline of forest carbon sequestration. Conservation and sustainable management can help to ensure our forests continue to remove carbon from the atmosphere while also improving soil and water quality, reducing wildfire risk, and otherwise managing forests to be more resilient in the fact of climate change. The Administration is working to identify new approaches to protect and restore our forests, as well as other critical landscapes including grasslands and wetlands, in the face of a changing climate.

V. Leading at the Federal Level

Leading in Clean Energy: President Obama believes that the federal government must be a leader in clean energy and energy efficiency. Under the Obama Administration, federal agencies have reduced greenhouse gas emissions by more than 15 percent – the equivalent of permanently taking 1.5 million cars off the road. To build on this record, the Administration is establishing a new goal: The federal government will consume 20 percent of its electricity from renewable sources by 2020 – more than double the current goal of 7.5 percent. In addition, the federal government will continue to pursue greater energy efficiency that reduces greenhouse gas emissions and saves taxpayer dollars.

Federal Government Leadership in Energy Efficiency: On December 2, 2011, President Obama signed a memorandum entitled "Implementation of Energy Savings Projects and Performance-Based Contracting for Energy Savings," challenging federal agencies, in support of the Better Buildings Challenge, to enter into $2 billion worth of performance-based contracts within two years. Performance contracts drive economic development, utilize private sector innovation, and increase efficiency at minimum costs to the taxpayer, while also providing long- term savings in energy costs. Federal agencies have committed to a pipeline of nearly $2.3 billion from over 300 reported projects. In coming months, the Administration will take a number of actions to strengthen efforts to promote energy efficiency, including through performance contracting. For example, in order to increase access to capital markets for investments in energy efficiency, the Administration will initiate a partnership with the private sector to work towards a standardized contract to finance federal investments in energy efficiency. Going forward, agencies will also work together to synchronize building codes – leveraging those policies to improve the efficiency of federally owned and supported building stock. Finally, the Administration will leverage the "Green Button" standard – which aggregates energy data in a secure, easy to use format – within federal facilities to increase their ability to manage energy consumption, reduce greenhouse gas emissions, and meet sustainability goals.

Prepare the United States for the Impacts of Climate Change

As we act to curb the greenhouse gas pollution that is driving climate change, we must also prepare for the impacts that are too late to avoid. Across America, states, cities, and communities are taking steps to protect themselves by updating building codes, adjusting the way they manage natural resources, investing in more resilient infrastructure, and planning for rapid recovery from damages that nonetheless occur. The federal government has an important role to play in supporting community-based preparedness and resilience efforts, establishing policies that promote preparedness, protecting critical infrastructure and public resources, supporting science and research germane to preparedness and resilience, and ensuring that federal operations and facilities continue to protect and serve citizens in a changing climate.

The Obama Administration has been working to strengthen America's climate resilience since its earliest days. Shortly after coming into office, President Obama established an Interagency Climate Change Adaptation Task Force and, in October 2009, the President signed an Executive Order directing it to recommend ways federal policies and programs can better prepare the Nation for change. In May 2010, the Task Force hosted the first National Climate Adaptation Summit, convening local and regional stakeholders and decision-makers to identify challenges and opportunities for collaborative action.

In February 2013, federal agencies released Climate Change Adaptation Plans for the first time, outlining strategies to protect their operations, missions, and programs from the effects of climate change. The Department of Transportation, for example, is developing guidance for incorporating climate change and extreme weather event considerations into coastal highway projects, and the Department of Homeland Security is evaluating the challenges of changing conditions in the Arctic and along our Nation's borders. Agencies have also partnered with communities through targeted grant and technical-assistance programs—for example, the Environmental Protection Agency is working with low-lying communities in North Carolina to assess the vulnerability of infrastructure investments to sea level rise and identify solutions to reduce risks. And the Administration has continued, through the U.S. Global Change Research Program, to support science and monitoring to expand our understanding of climate change and its impacts.

Going forward, the Administration will expand these efforts into three major, interrelated initiatives to better prepare America for the impacts of climate change:

I. Building Stronger and Safer Communities and Infrastructure

By necessity, many states, cities, and communities are already planning and preparing for the impacts of climate change. Hospitals must build capacity to serve patients during more frequent heat waves, and urban planners must plan for the severe storms that infrastructure will need to withstand. Promoting on-the-ground planning and resilient infrastructure will be at the core of our work to strengthen America's communities. Specific actions will include:

Directing Agencies to Support Climate-Resilient Investment: The President will direct federal agencies to identify and remove barriers to making climate-resilient investments; identify and remove counterproductive policies that increase vulnerabilities; and encourage and support smarter, more resilient investments, including through agency grants, technical assistance, and other programs, in sectors from transportation and water management to conservation and disaster relief. Agencies will also be directed to ensure that climate risk-management considerations are fully integrated into federal infrastructure and natural resource management planning. To begin meeting this challenge, the Environmental Protection Agency is committing to integrate considerations of climate change impacts and adaptive measures into major programs, including its Clean Water and Drinking Water State Revolving Funds and grants for brownfields cleanup, and the Department of Housing and Urban Development is already requiring grant recipients in the Hurricane Sandy–affected region to take sea-level rise into account.

Establishing a State, Local, and Tribal Leaders Task Force on Climate Preparedness: To help agencies meet the above directive and to enhance local efforts to protect communities, the President will establish a short-term task force of state, local, and tribal officials to advise on key actions the federal government can take to better support local preparedness and resilience-building efforts. The task force will provide recommendations on removing barriers to resilient investments, modernizing grant and loan programs to better support local efforts, and developing information and tools to better serve communities.

Supporting Communities as they Prepare for Climate Impacts: Federal agencies will continue to provide targeted support and assistance to help communities prepare for climate- change impacts. For example, throughout 2013, the Department of Transportation's Federal Highway Administration is working with 19 state and regional partners and other federal agencies to test approaches for assessing local transportation infrastructure vulnerability to climate change and extreme weather and for improving resilience. The Administration will continue to assist tribal communities on preparedness through the Bureau of Indian Affairs, including through pilot projects and by supporting participation in federal initiatives that assess climate change vulnerabilities and develop regional solutions. Through annual federal agency "Environmental Justice Progress Reports," the Administration will continue to identify innovative ways to help our most vulnerable communities prepare for and recover from the impacts of climate change. The importance of critical infrastructure independence was brought home in the Sandy response. The Federal Emergency Management Agency and the Department of Energy are working with the private sector to address simultaneous restoration of electricity and fuels supply.

Boosting the Resilience of Buildings and Infrastructure: The National Institute of Standards and Technology will convene a panel on disaster-resilience standards to develop a comprehensive, community-based resilience framework and provide guidelines for consistently safe buildings and infrastructure – products that can inform the development of private-sector standards and codes. In addition, building on federal agencies' "Climate Change Adaptation Plans," the Administration will continue efforts to increase the resilience of federal facilities and infrastructure. The Department of Defense, for example, is assessing the relative vulnerability of its coastal facilities to climate change. In addition, the President's FY 2014 Budget proposes $200 million through the Transportation Leadership Awards program for Climate Ready Infrastructure in communities that build enhanced preparedness into their planning efforts, and that have proposed or are ready to break ground on infrastructure projects, including transit and rail, to improve resilience.

Rebuilding and Learning from Hurricane Sandy: In August 2013, President Obama's Hurricane Sandy Rebuilding Task Force will deliver to the President a rebuilding strategy to be implemented in Sandy-affected regions and establishing precedents that can be followed elsewhere. The Task Force and federal agencies are also piloting new ways to support resilience in the Sandy-affected region; the Task Force, for example, is hosting a regional "Rebuilding by Design" competition to generate innovative solutions to enhance resilience. In the transportation sector, the Department of Transportation's Federal Transit Administration (FTA) is dedicating $5.7 billion to four of the area's most impacted transit agencies, of which $1.3 billion will be allocated to locally prioritized projects to make transit systems more resilient to future disasters. FTA will also develop a competitive process for additional funding to identify and support larger, stand-alone resilience projects in the impacted region. To build coastal resilience, the Department of the Interior will launch a $100 million competitive grant program to foster partnerships and promote resilient natural systems while enhancing green spaces and wildlife habitat near urban populations. An additional $250 million will be allocated to support projects for coastal restoration and resilience across the region. Finally, with partners, the U.S. Army Corps of Engineers is conducting a $20 million study to identify strategies to reduce the vulnerability of Sandy-affected coastal communities to future large-scale flood and storm events, and the National Oceanic and Atmospheric Administration will strengthen long-term coastal observations and provide technical assistance to coastal communities.

II. Protecting our Economy and Natural Resources

Climate change is affecting nearly every aspect of our society, from agriculture and tourism to the health and safety of our citizens and natural resources. To help protect critical sectors, while also targeting hazards that cut across sectors and regions, the Administration will mount a set of sector- and hazard-specific efforts to protect our country's vital assets, to include:

Identifying Vulnerabilities of Key Sectors to Climate Change: The Department of Energy will soon release an assessment of climate-change impacts on the energy sector, including power-plant disruptions due to drought and the disruption of fuel supplies during severe storms, as well as potential opportunities to make our energy infrastructure more resilient to these risks. In 2013, the Department of Agriculture and Department of the Interior released several studies outlining the challenges a changing climate poses for America's agricultural enterprise, forests, water supply, wildlife, and public lands. This year and next, federal agencies will report on the impacts of climate change on other key sectors and strategies to address them, with priority efforts focusing on health, transportation, food supplies, oceans, and coastal communities.

Promoting Resilience in the Health Sector: The Department of Health and Human Services will launch an effort to create sustainable and resilient hospitals in the face of climate change. Through a public-private partnership with the healthcare industry, it will identify best practices and provide guidance on affordable measures to ensure that our medical system is resilient to climate impacts. It will also collaborate with partner agencies to share best practices among federal health facilities. And, building on lessons from pilot projects underway in 16 states, it will help train public-health professionals and community leaders to prepare their communities for the health consequences of climate change, including through effective communication of health risks and resilience measures.

Promoting Insurance Leadership for Climate Safety: Recognizing the critical role that the private sector plays in insuring assets and enabling rapid recovery after disasters, the Administration will convene representatives from the insurance industry and other stakeholders to explore best practices for private and public insurers to manage their own processes and investments to account for climate change risks and incentivize policy holders to take steps to reduce their exposure to these risks.

Conserving Land and Water Resources: America's ecosystems are critical to our nation's economy and the lives and health of our citizens. These natural resources can also help ameliorate the impacts of climate change, if they are properly protected.

The Administration has invested significantly in conserving relevant ecosystems, including working with Gulf State partners after the Deepwater Horizon spill to enhance barrier islands and marshes that protect communities from severe storms. The Administration is also implementing climate-adaptation strategies that promote resilience in fish and wildlife populations, forests and other plant communities, freshwater resources, and the ocean. Building on these efforts, the President is also directing federal agencies to identify and evaluate additional approaches to improve our natural defenses against extreme weather, protect biodiversity and conserve natural resources in the face of a changing climate, and manage our public lands and natural systems to store more carbon.

Maintaining Agricultural Sustainability: Building on the existing network of federal climate- science research and action centers, the Department of Agriculture is creating seven new Regional Climate Hubs to deliver tailored, science-based knowledge to farmers, ranchers, and forest landowners. These hubs will work with universities and other partners, including the Department of the Interior and the National Oceanic and Atmospheric Administration, to support climate resilience. Its Natural Resources Conservation Service and the Department of the Interior's Bureau of Reclamation are also providing grants and technical support to agricultural water users for more water-efficient practices in the face of drought and long-term climate change.

Managing Drought: Leveraging the work of the National Disaster Recovery Framework for drought, the Administration will launch a cross-agency National Drought Resilience Partnership as a "front door" for communities seeking help to prepare for future droughts and reduce drought impacts. By linking information (monitoring, forecasts, outlooks, and early warnings) with drought preparedness and longer-term resilience strategies in critical sectors, this effort will help communities manage drought-related risks.

Reducing Wildfire Risks: With tribes, states, and local governments as partners, the Administration has worked to make landscapes more resistant to wildfires, which are exacerbated by heat and drought conditions resulting from climate change.

Federal agencies will expand and prioritize forest and rangeland restoration efforts in order to make natural areas and communities less vulnerable to catastrophic fire. The Department of the Interior and Department of Agriculture, for example, are launching a Western Watershed Enhancement Partnership – a pilot effort in five western states to reduce wildfire risk by removing extra brush and other flammable vegetation around critical areas such as water reservoirs.

Preparing for Future Floods: To ensure that projects funded with taxpayer dollars last as long as intended, federal agencies will update their flood-risk reduction standards for federally funded projects to reflect a consistent approach that accounts for sea-level rise and other factors affecting flood risks.

This effort will incorporate the most recent science on expected rates of sea-level rise (which vary by region) and build on work done by the Hurricane Sandy Rebuilding Task Force, which announced in April 2013 that all federally funded Sandy-related rebuilding projects must meet a consistent flood risk reduction standard that takes into account increased risk from extreme weather events, sea-level rise, and other impacts of climate change.

III. Using Sound Science to Manage Climate Impacts

Scientific data and insights are essential to help government officials, communities, and businesses better understand and manage the risks associated with climate change. The Administration will continue to lead in advancing the science of climate measurement and adaptation and the development of tools for climate-relevant decision-making by focusing on increasing the availability, accessibility, and utility of relevant scientific tools and information. Specific actions will include:

Developing Actionable Climate Science: The President's Fiscal Year 2014 Budget provides more than $2.7 billion, largely through the 13-agency U.S. Global Change Research Program, to increase understanding of climate-change impacts, establish a public-private partnership to explore risk and catastrophe modeling, and develop the information and tools needed by decision-makers to respond to both long-term climate change impacts and near-term effects of extreme weather.

Assessing Climate-Change Impacts in the United States: In the spring of 2014, the Obama Administration will release the third U.S. National Climate Assessment, highlighting new advances in our understanding of climate-change impacts across all regions of the United States and on critical sectors of the economy, including transportation, energy, agriculture, and ecosystems and biodiversity. For the first time, the National Climate Assessment will focus not only on dissemination of scientific information but also on translating scientific insights into practical, useable knowledge that can help decision-makers anticipate and prepare for specific climate-change impacts.

Launching a Climate Data Initiative: Consistent with the President's May 2013 Executive Order on Open Data – and recognizing that freely available open government data can fuel entrepreneurship, innovation, scientific discovery, and public benefits – the Administration is launching a Climate Data Initiative to leverage extensive federal climate-relevant data to stimulate innovation and private-sector entrepreneurship in support of national climate-change preparedness.

Providing a Toolkit for Climate Resilience: Federal agencies will create a virtual climate- resilience toolkit that centralizes access to data-driven resilience tools, services, and best practices, including those developed through the Climate Data Initiative. The toolkit will provide easy access to existing resources as well as new tools, including: interactive sea-level rise maps and a sea-level-rise calculator to aid post-Sandy rebuilding in New York and New Jersey, new NOAA storm surge models and interactive maps from the National Oceanic and Atmospheric Administration that provide risk information by combining tidal data, projected sea levels and storm wave heights, a web-based tool that will allow developers to integrate NASA climate imagery into websites and mobile apps, access to the U.S. Geological Survey's "visualization tool" to assess the amount of carbon absorbed by landscapes, and a Stormwater Calculator and Climate Assessment Tool developed to help local governments assess stormwater-control measures under different precipitation and temperature scenarios.

Lead International Efforts to Address Global Climate Change

The Obama Administration is working to build on the actions that it is taking domestically to achieve significant global greenhouse gas emission reductions and enhance climate preparedness through major international initiatives focused on spurring concrete action, including bilateral initiatives with China, India, and other major emitting countries. These initiatives not only serve to support the efforts of the United States and others to achieve our goals for 2020, but also will help us move beyond those and bend the post-2020 global emissions trajectory further. As a key part of this effort, we are also working intensively to forge global responses to climate change through a number of important international negotiations, including the United Nations Framework Convention on Climate Change.

I. Working with Other Countries to Take Action to Address Climate Change

Enhancing Multilateral Engagement with Major Economies: In 2009, President Obama launched the Major Economies Forum on Energy and Climate, a high-level forum that brings together 17 countries that account for approximately 75 percent of global greenhouse gas emissions, in order to support the international climate negotiations and spur cooperative action to combat climate change. The Forum has been successful on both fronts – having contributed significantly to progress in the broader negotiations while also launching the Clean Energy Ministerial to catalyze the development and deployment of clean energy and efficiency solutions. We are proposing that the Forum build on these efforts by launching a major initiative this year focused on further accelerating efficiency gains in the buildings sector, which accounts for approximately one-third of global carbon pollutions from the energy sector.

Expanding Bilateral Cooperation with Major Emerging Economies: From the outset, the Obama Administration has sought to intensify bilateral climate cooperation with key major emerging economies, through initiatives like the U.S.-China Clean Energy Research Center, the U.S.-India Partnership to Advance Clean Energy, and the Strategic Energy Dialogue with Brazil.

We will be building on these successes and finding new areas for cooperation in the second term, and we are already making progress: Just this month, President Obama and President Xi Jinping of China reached an historic agreement at their first summit to work to use the expertise and institutions of the Montreal Protocol to phase down the consumption and production of HFCs, a highly potent greenhouse gas. The impact of phasing out HFCs by 2050 would be equivalent to the elimination of two years' worth of greenhouse gas emissions from all sources.

Combatting Short-Lived Climate Pollutants: Pollutants such as methane, black carbon, and many HFCs are relatively short-lived in the atmosphere, but have more potent greenhouse effects than carbon dioxide. In February 2012, the United States launched the Climate and Clean Air Coalition to Reduce Short-Lived Climate Pollution, which has grown to include more than 30 country partners and other key partners such as the World Bank and the U.N. Environment Programme. Major efforts include reducing methane and black carbon from waste and landfills. We are also leading through the Global Methane Initiative, which works with 42 partner countries and an extensive network of over 1,100 private sector participants to reduce methane emissions.

Reducing Emissions from Deforestation and Forest Degradation: Greenhouse gas emissions from deforestation, agriculture, and other land use constitute approximately one-third of global emissions. In some developing countries, as much as 80 percent of these emissions come from the land sector. To meet this challenge, the Obama Administration is working with partner countries to put in place the systems and institutions necessary to significantly reduce global land-use-related emissions, creating new models for rural development that generate climate benefits, while conserving biodiversity, protecting watersheds, and improving livelihoods.

In 2012 alone, the U.S. Agency for International Development's bilateral and regional forestry programs contributed to reducing more than 140 million tons of carbon dioxide emissions, including through support for multilateral initiatives such as the Forest Investment Program and the Forest Carbon Partnership Facility. In Indonesia, the Millennium Challenge Corporation is funding a five-year "Green Prosperity" program that supports environmentally sustainable, low carbon economic development in select districts.

The Obama Administration is also working to address agriculture-driven deforestation through initiatives such as the Tropical Forest Alliance 2020, which brings together governments, the private sector, and civil society to reduce tropical deforestation related to key agricultural commodities, which we will build upon.

Expanding Clean Energy Use and Cut Energy Waste: Roughly 84 percent of current carbon dioxide emissions are energy-related and about 65 percent of all greenhouse gas emissions can be attributed to energy supply and energy use. The Obama Administration has promoted the expansion of renewable, clean, and efficient energy sources and technologies worldwide through:

- Financing and regulatory support for renewable and clean energy projects
- Actions to promote fuel switching from oil and coal to natural gas or renewables
- Support for the safe and secure use of nuclear power
- Cooperation on clean coal technologies
- Programs to improve and disseminate energy efficient technologies

In the past three years we have reached agreements with more than 20 countries around the world, including Mexico, South Africa, and Indonesia, to support low emission development strategies that help countries to identify the best ways to reduce greenhouse gas emissions while growing their economies. Among the many initiatives that we have launched are:

- The U.S. Africa Clean Energy Finance Initiative, which aligns grant-based assistance with project planning expertise from the U.S. Trade and Development Agency and financing and risk mitigation tools from the U.S. Overseas Private Investment Corporation to unlock up to $1 billion in clean energy financing.
- The U.S.-Asia Pacific Comprehensive Energy Partnership, which has identified $6 billion in U.S. export credit and government financing to promote clean energy development in the Asia-Pacific region.

Looking ahead, we will target these and other resources towards greater penetration of renewables in the global energy mix on both a small and large scale, including through our participation in the Sustainable Energy for All Initiative and accelerating the commercialization of renewable mini-grids. These efforts include:

- *Natural Gas.* Burning natural gas is about one-half as carbon-intensive as coal, which can make it a critical "bridge fuel" for many countries as the world transitions to even cleaner sources of energy. Toward that end, the Obama Administration is partnering with states and private companies to exchange lessons learned with our international partners on responsible development of natural gas resources. We have launched the Unconventional Gas Technical Engagement Program to share best practices on issues such as water management, methane emissions, air quality, permitting, contracting, and pricing to help increase global gas supplies and facilitate development of the associated infrastructure that brings them to market. Going forward, we will promote fuel-switching from coal to gas for electricity production and encourage the development of a global market for gas. Since heavy-duty vehicles are expected to account for 40 percent of increased oil use through 2030, we will encourage the adoption of heavy duty natural gas vehicles as well.
- *Nuclear Power.* The United States will continue to promote the safe and secure use of nuclear power worldwide through a variety of bilateral and multilateral engagements. For example, the U.S. Nuclear Regulatory Commission advises international partners on safety and regulatory best practices, and the Department of Energy works with international partners on research and development, nuclear waste and storage, training, regulations, quality control, and comprehensive fuel leasing options. Going forward, we will expand these efforts to promote nuclear energy generation consistent with maximizing safety and nonproliferation goals.
- *Clean Coal.* The United States works with China, India, and other countries that currently rely heavily on coal for power generation to advance the development and deployment of clean coal technologies. In addition, the U.S. leads the Carbon Sequestration Leadership Forum, which engages 23 other countries and economies on carbon capture and sequestration technologies. Going forward, we will

continue to use these bilateral and multilateral efforts to promote clean coal technologies.

- *Energy Efficiency.* The Obama Administration has aggressively promoted energy efficiency through the Clean Energy Ministerial and key bilateral programs. The cost- effective opportunities are enormous: The Ministerial' s Super-Efficient Equipment and Appliance Deployment Initiative and its Global Superior Energy Performance Partnership are helping to accelerate the global adoption of standards and practices that would cut energy waste equivalent to more than 650 mid-size power plants by 2030. We will work to expand these efforts focusing on several critical areas, including: improving building efficiency, reducing energy consumption at water and wastewater treatment facilities, and expanding global appliance standards.

Negotiating Global Free Trade in Environmental Goods and Services: The U.S. will work with trading partners to launch negotiations at the World Trade Organization towards global free trade in environmental goods, including clean energy technologies such as solar, wind, hydro and geothermal. The U.S. will build on the consensus it recently forged among the 21 Asia-Pacific Economic Cooperation (APEC) economies in this area. In 2011, APEC economies agreed to reduce tariffs to 5 percent or less by 2015 on a negotiated list of 54 environmental goods. The APEC list will serve as a foundation for a global agreement in the WTO, with participating countries expanding the scope by adding products of interest. Over the next year, we will work towards securing participation of countries which account for 90 percent of global trade in environmental goods, representing roughly $481 billion in annual environmental goods trade. We will also work in the Trade in Services Agreement negotiations towards achieving free trade in environmental services.

Phasing Out Subsidies that Encourage Wasteful Consumption of Fossil Fuels: The International Energy Agency estimates that the phase-out of fossil fuel subsidies – which amount to more than $500 billion annually – would lead to a 10 percent reduction in greenhouse gas emissions below business as usual by 2050. At the 2009 G-20 meeting in Pittsburgh, the United States successfully advocated for a commitment to phase out these subsidies, and we have since won similar commitments in other fora such as APEC. President Obama is calling for the elimination of U.S. fossil fuel tax subsidies in his

Fiscal Year (FY) 2014 budget, and we will continue to collaborate with partners around the world toward this goal.

Leading Global Sector Public Financing Towards Cleaner Energy: Under this Administration, the United States has successfully mobilized billions of dollars for clean energy investments in developing countries, helping to accelerate their transition to a green, low-carbon economy. Building on these successes, the President calls for an end to U.S. government support for public financing of new coal plants overseas, except for (a) the most efficient coal technology available in the world's poorest countries in cases where no other economically feasible alternative exists, or (b) facilities deploying carbon capture and sequestration technologies. As part of this new commitment, we will work actively to secure the agreement of other countries and the multilateral development banks to adopt similar policies as soon as possible.

Strengthening Global Resilience to Climate Change: Failing to prepare adequately for the impacts of climate change that can no longer be avoided will put millions of people at risk, jeopardizing important development gains, and increasing the security risks that stem from climate change. That is why the Obama Administration has made historic investments in bolstering the capacity of countries to respond to climate-change risks. Going forward, we will continue to:

- Strengthen government and local community planning and response capacities, such as by increasing water storage and water use efficiency to cope with the increased variability in water supply
- Develop innovative financial risk management tools such as index insurance to help smallholder farmers and pastoralists manage risk associated with changing rainfall patterns and drought
- Distribute drought-resistant seeds and promote management practices that increase farmers' ability to cope with climate impacts.

Mobilizing Climate Finance: International climate finance is an important tool in our efforts to promote low-emissions, climate-resilient development. We have fulfilled our joint developed country commitment from the Copenhagen Accord to provide approximately $30 billion of climate assistance to developing countries over FY 2010-FY 2012. The United States contributed approximately $7.5 billion to this effort over the three year period. Going forward, we will seek to build on this progress as well as focus our efforts on combining our public resources with smart policies to mobilize much larger flows of private investment in low-emissions and climate resilient infrastructure.

II. Leading Efforts to Address Climate Change through International Negotiations

The United States has made historic progress in the international climate negotiations during the past four years. At the Copenhagen Conference of the United Nations Framework Convention on Climate Change (UNFCCC) in 2009, President Obama and other world leaders agreed for the first time that all major countries, whether developed or developing, would implement targets or actions to limit greenhouse emissions, and do so under a new regime of international transparency. And in 2011, at the year-end climate meeting in Durban, we achieved another breakthrough: Countries agreed to negotiate a new agreement by the end of 2015 that would have equal legal force and be applicable to all countries in the period after 2020. This was an important step beyond the previous legal agreement, the Kyoto Protocol, whose core obligations applied to developed countries, not to China, India, Brazil or other emerging countries.

The 2015 climate conference is slated to play a critical role in defining a post-2020 trajectory. We will be seeking an agreement that is ambitious, inclusive and flexible. It needs to be ambitious to meet the scale of the challenge facing us. It needs to be inclusive because there is no way to meet that challenge unless all countries step up and play their part. And it needs to be flexible because there are many differently situated parties with their own needs and imperatives, and those differences will have to be accommodated in smart, practical ways.

At the same time as we work toward this outcome in the UNFCCC context, we are making progress in a variety of other important negotiations as well. At the Montreal Protocol, we are leading efforts in support of an amendment that would phase down HFCs; at the International Maritime Organization, we have agreed to and are now implementing the first-ever sector-wide, internationally applicable energy efficiency standards; and at the International Civil Aviation Organization, we have ambitious aspirational emissions and energy efficiency targets and are working towards agreement to develop a comprehensive global approach.

INDEX

C

D

E

H

I

J

K

L

M

N

O

T

U

V

W